Container Port Production and Economic Efficiency

Container Port Production and Economic Efficiency

Teng-Fei Wang,
Kevin Cullinane
and
Dong-Wook Song

First published in 2005 by
PALGRAVE MACMILLAN
Houndmills, Basingstoke, Hampshire RG21 6XS and
175 Fifth Avenue, New York, N.Y. 10010
Companies and representatives throughout the world.

PALGRAVE MACMILLAN is the global academic imprint of the Palgrave Macmillan division of St. Martin's Press, LLC and of Palgrave Macmillan Ltd. Macmillan® is a registered trademark in the United States, United Kingdom and other countries. Palgrave is a registered trademark in the European Union and other countries.

ISBN-13: 978–1–4039–4772–7 hardback
ISBN-10: 1–4039–4772–4 hardback

This book is printed on paper suitable for recycling and made from fully managed and sustained forest sources.

A catalogue record for this book is available from the British Library.

Library of Congress Cataloging-in-Publication Data

Wang, Teng-Fei.
    Container port production and economic efficiency / Teng-Fei Wang, Kevin Cullinane, and DongWook Song.
        p. cm.
    Includes bibliographical references and index.
    ISBN 1–4039–4772–4
        1. Container ports – Management. 2. Container ports – Cost effectiveness. 3. Container ports – Economic aspects. 4. Unitized cargo systems – Management. I. Cullinane, Kevin. II. Song, Dong-Wook. III. Title.

HE551.W36 2005
387.1'64'068—dc22   2005042929

10 9 8 7 6 5 4 3 2 1
14 13 12 11 10 09 08 07 06 05

Printed and bound in Great Britain by
Antony Rowe Ltd, Chippenham and Eastbourne

Dr Teng-Fei Wang dedicates this book to his father, Mingcheng.

Prof. Kevin Cullinane dedicates this book to the memory of Brian Yolland; a true shipping man, a natural orator, a worthy mentor and a good friend. Long lost, but long remembered.

Dr Dong-Wook Song dedicates this book to his beloved family members, Sung-Hee, Jee-Young and Jee-Hoon, for love and to the Lord for the good.

Contents

List of Tables		xi
List of Figures		xii
Glossary of Symbols and Abbreviations		xiv
Acknowledgements		xvi
Foreword		xvii

1 Introduction | | **1**
1.1 The importance of the container port industry | 1
1.2 Performance measurement, productivity and efficiency | 2
1.3 Measuring performance in the container port industry | 8
1.4 Problem statement | 9
1.5 Objectives | 11
1.6 Organisation of the book | 11

2 The Economic Theory of Container Port Production | **13**
2.1 Introduction | 13
2.2 The economic functions of ports and their evolution | 14
2.3 The container port industry: the challenge of complexity | 16
2.4 The challenge from other market players | 18
2.5 Traditional theories of industrial market structure | 21
2.6 Port competition | 22
   2.6.1 Inter-port competition | 22
   2.6.2 Intra-port competition | 27
2.7 Summary | 32

3 Alternative Approaches to Efficiency Measurement | **33**
3.1 Background | 33
3.2 Notation and symbols used | 37
3.3 Data envelopment analysis | 39
   3.3.1 Introduction to DEA | 39
   3.3.2 DEA approaches using cross-sectional data | 40
   3.3.3 DEA approaches using panel data | 45

3.4 Free disposal hull analysis 49
3.5 Econometric approaches using
cross-sectional data 51
   3.5.1 Deterministic frontier econometric
approaches 51
   3.5.2 Stochastic frontier approaches 55
3.6 Econometric approaches using panel data 58
3.7 Empirical comparisons of alternative
approaches to efficiency estimation 61
3.8 Performance measurement and efficiency
analysis in the container port industry 63
   3.8.1 General performance measurement 63
   3.8.2 Efficiency measurement in the
port industry: the DEA approach 66
   3.8.3 Efficiency measurement in the port
industry: the SFA approach 69
   3.8.4 Lessons to be learned from applying
DEA and SFA to port efficiency measurement 72
3.9 Summary 75

4 Model Specification and Data **77**
4.1 Introduction 77
4.2 A procedure for efficiency measurement 77
4.3 Port objectives 79
4.4 Definition of variables 81
4.5 Model specification 87
4.6 Data collection 89
4.7 Summary 92

5 Empirical Results and Analysis **93**
5.1 Introduction 93
5.2 Cross-sectional data analysis 93
   5.2.1 Non-parametric frontier estimation: DEA
versus FDH 93
   5.2.2 Parametric frontier estimation: COLS versus SFA 98
   5.2.3 Comparison between non-parametric and
parametric frontier models using
cross-sectional data 102
5.3 Panel data analysis 103
   5.3.1 DEA panel analysis 103

5.3.2 Time-invariant efficiency analysis using
panel data 106
5.3.3 Time-varying efficiency analysis using
panel data 109
5.3.4 Comparison between non-parametric
and parametric frontier models using
panel data 110
5.4 Summary 113

**6 A Theory of Container Port Production and its
Empirical Validation** **116**
6.1 Introduction 116
6.2 Testing the hypotheses 116
6.3 Summary 136

7 Conclusions and Further Research **140**
7.1 Introduction 140
7.2 A summary of major research contributions 140
7.2.1 Contribution to container port
production theory 140
7.2.2 Contribution to methodological choice in
measuring efficiency 143
7.3 Recommendations for further research 144

Appendices **147**
Appendix 1 Terminal Efficiency of the CCR,
BCC and FDH Models 147
Appendix 2 Efficiency Yielded from Alternative
Parametric Approaches 149
Appendix 3 DEA–CCR Window Analysis for
Container Port Efficiency 151
Appendix 4 DEA–BCC Window Analysis for
Container Port Efficiency 158
Appendix 5 DEA–CCR Contemporaneous Analysis
of Container Port Efficiency 165
Appendix 6 DEA–BCC Contemporaneous Analysis of
Container Port Efficiency 166
Appendix 7 DEA–CCR Intertemporal Analysis of
Container Port Efficiency 167
Appendix 8 DEA–BCC Intertemporal Analysis of
Container Port Efficiency 168

Appendix 9   Technical-invariant Efficiency
Estimated by Alternative Models   169

Appendix 10   Time-varying Efficiency by MLE   170

References   171

Index   183

List of Tables

2.1	Port facilities and services	14
2.2	The development of ports	15
2.3	Linkages between economic globalisation and the supply of transportation resources	19
2.4	Port function matrix	28
3.1	Taxonomy of frontier methodologies	36
3.2	Single input and single output	40
3.3	Efficiency estimates by maximum likelihood	57
3.4	Summary of performance indicators suggested by UNCTAD	64
3.5	The application of DEA to ports	67
3.6	The application of SFA to ports	70
4.1	Summary statistics for sample 1	90
4.2	Summary statistics for sample 2	91
5.1	Summary of the terminal efficiency estimates of the CCR, BCC and FDH models	94
5.2	Summary results on numbers of efficient terminals with DEA models and the FDH model	96
5.3	Frontier production functions of container terminals	100
5.4	Summary of the efficiency yielded from alternative parametric approaches	101
5.5	Spearman's rank order correlation coefficient	103
5.6	Frontier production function of container ports	108
5.7	Spearman's rank order correlation coefficient	109
5.8	Frontier production function of container ports	109
6.1	Relationship between production size and efficiency	118
6.2	Relationship between scale of production and fluctuations in efficiency	119
6.3	Markets and inter-port competition	124
6.4	Relationship between inter-port competition and efficiency	130
6.5	Ownership status of the sample container ports	130
6.6	Terminal efficiency of the CCR, BCC and FDH models	137

List of Figures

1.1   DMU and homogeneous units   2
1.2   Performance measures and organisational development   3
1.3   Illustration of efficiency and productivity   5
1.4   Comparison between technical and allocative efficiency   7
2.1   Privatisation and port performance   30
3.1   Farrell's measurement of efficiency   34
3.2   Alternative approaches to efficiency
     measurement derived from Farrell   35
3.3   Comparison of efficiencies of container terminals   41
3.4   Comparison of efficiencies of container terminals   43
3.5   Conceptual illustration of contemporaneous,
     intertemporal and window analyses   47
3.6   Non-parametric deterministic frontiers   49
3.7   Alternative panel data approaches to
     efficiency measurement   62
4.1   A performance measuring system   78
4.2   Constraining influences on port management   81
5.1   Sample size and proportion of inefficient firms   97
5.2   Skewness of the OLS residuals for
     cross-sectional data   99
5.3   Comparison of efficiency yielded from
     various approaches   102
5.4   Year-by-year average efficiency for all container ports   105
5.5   Skewness of the OLS residuals for panel data   107
5.6   Comparison of average efficiency by
     alternative approaches   111
5.7   Comparison of the efficiency of Hong Kong
     container port yielded by alternative approaches   112
6.1   Overall level of concentration in the container
     port industry as measured by the HHI   121
6.2   Geographic locations of the world's top container ports   123
6.3   Hypothesised relationship between inter-port
     competition and efficiency   125
6.4   Relationship between inter-port competition
     and efficiency with reference to cross-sectional data   126

6.5   Relationship between inter-port competition
      and efficiency with reference to cross-sectional data   127
6.6   Relationship between inter-port competition
      and efficiency with reference to panel data
      (Contemporaneous analysis)   128
6.7   Relationship between inter-port competition
      and efficiency with reference to panel data
      (Intertemporal analysis)   129
6.8   Ownership classification of the container ports
      in the sample   131
6.9   Ownership versus efficiency   132
6.10 Ownership versus efficiency   133
6.11 Ownership versus efficiency with reference to
      container ports in China   134
6.12 Ownership versus efficiency with reference to
      container ports in China   135
6.13 Efficiency and variation of different terminals at
      individual ports   138

Glossary of Symbols and Abbreviations

AE	Allocative Efficiency
CRS	Constant Returns to Scale
DEA	Data Envelopment Analysis
DMU	Decision Making Unit
FDH	Free Disposal Hull
FP	Fractional Programming
iid	independently and identically distributed
k	the kth firm
K	the number of firms
LP	Linear Programming
m	the mth input variable for a firm
M	the number of input variables for a firm
MLE	Maximum Likelihood Estimation
n	the nth output variable for a firm
N	the number of output variables for a firm
OLS	Ordinary Least Squares
SE	Scale Efficiency
SFA	Stochastic Frontier Analysis
t	the tth time period
T	the total number of time periods observed
TE	Technical Efficiency
TEU	Twenty-foot Equivalent Unit
u	$\ln U$
U	the value of technical efficiency
v	$\ln V$
V	the value of noise component
VRS	Variable Returns to Scale
w	the window width describing the time duration for the observations subsets
x	the amount of a factor used (input)
x_{mkt}	the mth input data of firm k at time t
X	the matrix of input variables
y	the amount of the product produced (output)

y_{nkt} the nth output data of firm k at time t

Y the matrix of output variables

Z time-invariant observable variables which differs across units

$\phi(\cdot)$ a density function

β a parameter vector of input variable, $\beta = (\beta_1, \beta_2, \ldots, \beta_M)$

$\Phi(\cdot)$ the standard normal cumulative distribution

Acknowledgements

The authors are grateful to a number of individuals who have contributed to bringing this book to fruition. The work of the editorial staff at Palgrave-Macmillan has been unstinting. In particular, a great debt is owed to Jacky Kippenberger and to Katie Button, whose help, advice and feedback provided us with the much-needed motivation to meet an ambitious deadline.

All the authors are grateful to the Hong Kong Polytechnic University for providing the initial opportunity which allowed us to undertake this work. The support from staff in Hong Kong was invaluable and we would mention in particular the unparalleled contributions of Teresa Tong, Dr Ji Ping and Dr Josephine Khu.

At the University of Newcastle in the UK, Dr Teng-Fei Wang and Prof. Kevin Cullinane owe a debt of gratitude to the Head of the School of Marine Science and Technology, Prof. Atilla Incecik for having the vision to allow us the space, time and resources to allow us to conduct the research that has culminated in this book. Finally, thanks should also go to Dr Sharon Cullinane for not only showing immense patience with one of the authors, but also for proof-reading the final work and offering valuable advice on structure, style and literacy.

The author and publishers have made every attempt to contact copyright holders. If any have inadvertently been overlooked, the appropriate arrangements will be made at the first opportunity.

Foreword

With the globalisation of the world economy, the container port industry is becoming increasingly important. This research is motivated by the contrast between the ever-mounting importance of the contemporary container port industry and the sparsity of scientific and in-depth research of the economic theories that underpin it. Despite the paramount contribution of the container port industry to globalisation and international trade, the relevance and appropriateness of many fundamental economic theories underpinning container port production are simply assumed and, in fact, deserve to be more thoroughly investigated.

This research is also motivated by the vital role played by efficiency measurement in any sort of production and the dearth of such studies in the container port industry. Traditional approaches are confined to partial measures of productivity and are not sophisticated enough either to reflect the complexity of contemporary container port production or to provide sufficient insights into the management or policy implications of efficiency estimates. In recent years, two leading approaches to measuring efficiency, Data Envelopment Analysis (DEA) and Stochastic Frontier Analysis (SFA), have been occasionally applied to ports or, more specifically, to the container port industry in order to measure their efficiency. However, previous research in this area is far from sufficient. Among other reasons for this, the existing corpus of research is either based on strong assumptions, or has ignored the great variety and diverse nature of the available data (such as cross-sectional or longitudinal data).

Against this background, this book contributes to the existing literature in two ways. First, the economic theories underpinning container port production (such as the relationship between ownership, competition and port efficiency) are not only analysed by applying traditional economic theory (in particular that of industrial organisation theory), but are also examined empirically by deriving scientific estimates of efficiency. Most work in this aspect is original and, potentially, makes an important contribution to the establishment of central government policy on port investment, policy and governance. Second, for the first time, comprehensive comparisons of alternative approaches to efficiency measurement are conducted for the container port industry. These approaches include the two most well known and commonly applied, Data Envelopment Analysis (DEA) and Stochastic Frontier

Analysis (SFA), as well as some other important alternatives, such as the Free Disposal Hull (FDH) method. In addition, consideration is given to the use of panel data and to random- and fixed-effects models. Due to the individual strengths and weaknesses associated with the various approaches to efficiency measurement, this sort of comparative study represents both a significant and necessary contribution to both the theoretical and empirical aspects of contemporary efficiency measurement.

1
Introduction

1.1 The importance of the container port industry

In its simplest form, 'Globalisation' refers to the increasing geographical scale of economic, social and political interactions. International trade and its integral activities of importing and exporting constitute the fundamental aspects of globalisation (Janelle and Beuthe, 1997).

Transportation plays a dual role as a proactive agent of globalisation on the one hand, and as a beneficiary of its development on the other. Together with enhanced telecommunications, the continued liberalisation of trade and efforts at international standardisation, transportation is regarded as one of the four cornerstones of globalisation (Janelle and Beuthe, 1997, 2002; Peters, 2001; Goulielmos and Pardali, 2002; Kumar and Hoffman, 2002). Among the alternative modes of freight transportation, shipping plays the dominant role; accounting for almost two-thirds of world trade in metric tonnes (UNCTAD, 2001). In particular, container transportation has been playing an increasingly important role in the process, largely because of the numerous technical and economic advantages it possesses over traditional methods of transportation.

Standing at the crucial interface of sea and inland transportation, the significance of the container port and its production capabilities cannot be ignored. Many activities are necessarily involved in determining a ship's turnaround time in port and the cost incurred by users of the port. This is especially so for the most obvious users such as shipping lines and shippers, but also for those that are not so immediately identifiable, such as the producers and final users of the products that comprise the cargoes carried.

1.2   Performance measurement, productivity and efficiency

Production can be simply defined as a process by which inputs are combined, transformed and turned into outputs (Case and Fair, 1999). It is a fundamental concept in economic theory. Figure 1.1 depicts the production of a group of K firms using M inputs to produce N outputs. The inputs can normally be generalised as natural resources such as land, human resources (labour) and man-made aids to further production such as tools and machinery (capital). Outputs, on the other hand, can be categorised into tangible products, including goods, and intangible products such as services. The production unit that transforms inputs into outputs is frequently referred to as either the 'Decision Making Unit (DMU)' in the literature on management science and as the 'firm' in economics (Greene, 1993). In this book, the terms 'DMU' and 'firm' can be considered as interchangeable, though the term 'firm' is used whenever possible. However, the use of the term 'firm' is not always appropriate (Coelli *et al.*, 1998, p. 1). For instance, when comparing the performance of a group of container terminals within the same port (i.e. a case of intra-port competition), the units under consideration might be *parts* of a firm rather than firms themselves. In such a case, the units are treated as different 'firms' in this book.

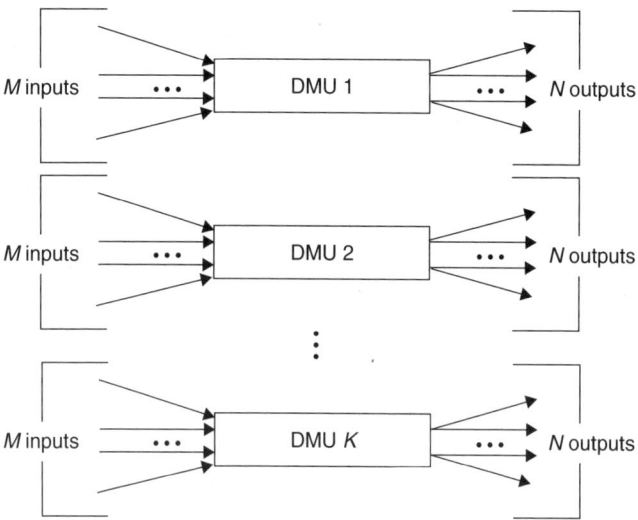

Figure 1.1   DMU and homogeneous units.
Source: Drawn by the authors.

Given the limited resources available in every possible example of 'production', it is important to study the quality of that production, including the efficient use of inputs and the maximisation of outputs. This has led to the important concept of 'performance', which is used to describe the quality of production. Dyson (2000) claims that the measurement of performance plays an essential role in evaluating production because, as shown in Figure 1.2, performance can define not only the current state of the system, but also its future. The measurement of performance helps move the system in the desired direction through the effect exerted by the behavioural responses to the performance measures that exist within the system. Misspecified performance measures, however, will lead to unintended consequences, with the system moving in the wrong direction.

Thanassoulis (2001) has identified the following information that can be obtained by measuring performance:

- the identification of good operating practices for dissemination;
- the most productive operating scale;
- the scope for efficiency savings in the use of resources and/or for augmenting output;
- the most suitable role model for an inefficient unit to emulate in order to improve its performance;
- the marginal rates of substitution between the factors of production; and
- the productivity change over time by each operating unit and by the most efficient of the operating units at each point in time.

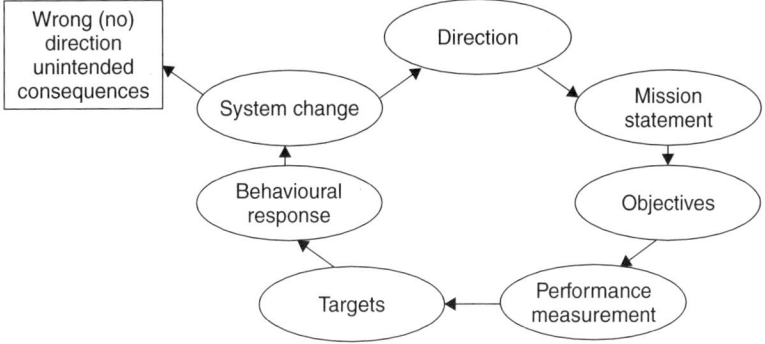

Figure 1.2 Performance measures and organisational development.

Source: Dyson (2000, p. 5). Copyright © Operational Research Society. Reproduced by kind permission.

Productivity and efficiency are the two most important concepts in measuring performance and are frequently used interchangeably. However, in accordance with the work of Coelli *et al.* (1998), for the purposes of this book, a distinction has been drawn between their respective definitions. The productivity of a firm in this book is defined either as the ratio of output to input or as *total factor productivity*. These correspond, respectively, to the situations where there is a single output and single input or where there are multiple outputs and inputs (Coelli *et al.*, 1998).

Both efficiency and productivity are important concepts in the field of traditional economics. In his influential and seminal paper, Leibenstein (1966) explicitly stated that: 'At the core of economics is the concept of efficiency.' Recently, Forsund and Sarafoglou (2002) claimed that 'Efficiency and productivity are core concepts of economics.' Lovell (1993, p. 5) has suggested two reasons why it is significant to measure efficiency and productivity: first, they are indicators of success by which production units can be evaluated and second, they enable us to explore hypotheses concerning the sources of efficiency and productivity differentials. Identifying these sources is essential to instituting public and private policies designed to improve performance. Moreover, macroperformance depends upon microperformance, and so the same reasoning applies to the study of the economic development and growth of nations.

Efficiency is a relative concept that can only be evaluated through a process of comparison or benchmarking. Efficiency comprises *technical efficiency*, *scale efficiency* and *allocative efficiency*. *Technical efficiency* is defined as relative productivity over time or space, or both. For instance, it can be divided into intra- and inter-firm measures of efficiency. The former involves measuring the firm's own production potential by computing its productivity level over time, relative to that of a firm-specific highest level of historic productivity. By contrast, the latter measures the performance of a particular firm relative to its best comparator counterpart(s) available in the industry (Lansink *et al.*, 2001). The concept of *technical efficiency* is also closely associated with two other important constructs; the 'production frontier' and the 'cost frontier' that appear within the context of theoretical economics. The former refers to the set of maximum outputs given different levels of input, while the latter indicates the set of minimum inputs given different levels of output. The production frontier reflects the current state of technology in an industry. Corresponding to the production or cost frontier, technical efficiency can be differentiated respectively as output- and input-oriented technical efficiencies; that is, the producer can either improve output(s) given the same level of input(s) or reduce the input(s) given the same level of output(s).

Scale efficiency relates to a possible divergence between actual and ideal production size. The ideal configuration coincides with the long-run competitive equilibrium, where production is characterised by constant returns to scale. A producer is scale-efficient if its choice of inputs and outputs is situated in that part of a frontier (either production or cost) that yields constant returns to scale; where this is not the case, it is deemed to be scale-inefficient (De Borger *et al.*, 2002).

Allocative efficiency focuses on the costs of production given that information on prices is available and that a behavioural assumption, such as cost minimisation or profit maximisation, can be properly established and, therefore, constitutes an appropriate assumption to make. It is clear that *allocative efficiency* is different from *technical* and *scale* efficiencies in that the former addresses issues such as costs or profits, while the latter only considers physical quantities and technical relationships. For instance, allocative efficiency in input selection occurs when a selection of inputs (e.g. materials, labour and capital) produces a given quantity of output at a minimum cost, given the prevailing prices of the inputs (Coelli *et al.*, 1998, p. 5).

The difference between technical efficiency, scale efficiency and productivity can be simply illustrated, as shown in Figure 1.3. Points *A*, *B* and *C* refer to three different firms. The productivity of point A is measured by the ratio *DA/OD* according to the definition of productivity in which the *x*-axis represents inputs and the *y*-axis denotes outputs.

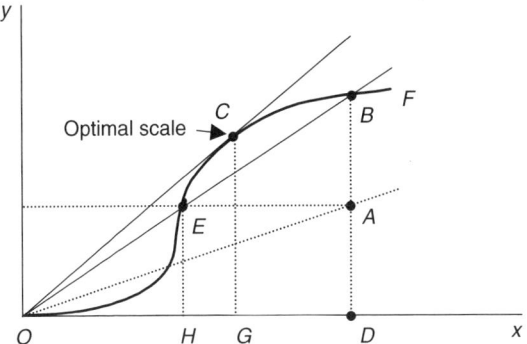

Figure 1.3 Illustration of efficiency and productivity.
Source: Derived from Coelli *et al.* (1998, p. 5).

The heavy curve *OF* in Figure 1.3 is the so-called production frontier. All of the points on the production frontier are technically efficient, while all of the points below or lying to the right of the efficient frontier are technically inefficient. Given the same input, it is quite clear that productivity can be further improved by moving from point *A* to point *B*. The new level of productivity is then given by *DB/OD*. Clearly, productivity can be represented, therefore, by the slope of the ray through the origin, which joins the specific point under study. The output-oriented technical efficiency of firm *A* can be measured by the ratio of the productivity of point *A* to that of point *B*, that is *(DA/OD)/(DB/OD) = DA/DB*, which indicates the ratio of the output actually attained to that which is potentially attainable, given a predetermined and set input level. Similarly, the input-oriented technical efficiency of firm *A* can be measured as *OH/OD*, which indicates the ratio of inputs needed to produce output *y*, relative to the input actually used to produce output *y*.

The ray through the origin and point *C* in Figure 1.3 is at a tangent to the production frontier and, hence, defines the point of maximum possible productivity.

The proliferation of attempts to measure efficiency through the application of the frontier approach can be attributed to an interest in the structure of efficient production technology, an interest in the divergence between an observed and an idealised production operation, and also to an interest in the concept of efficiency itself. Bauer (1990) suggests several reasons why the use of frontier models is becoming increasingly widespread:

- the notion of a frontier is consistent with the underlying economic theory of optimising behaviour;
- deviations from a frontier have a natural interpretation as a measure of the relative efficiency with which economic units pursue their technical or behavioural objectives; and
- information about the structure of the frontier and about the relative efficiency that firms (DMUs) can have on many policy applications.

The weakness of Figure 1.3 is that it is incapable of analysing *allocative efficiency*. This is overcome by Figure 1.4, which depicts a production process with two inputs, x_1 and x_2, that are combined to produce the same amount of output *y*. Making the assumption that all the points on the isoquant *uu* are technically efficient, then point *A* in Figure 1.4 represents a position where the firm is observed using (x_{1A}, x_{2A}) to produce *y*. The technical efficiency can then be measured by the

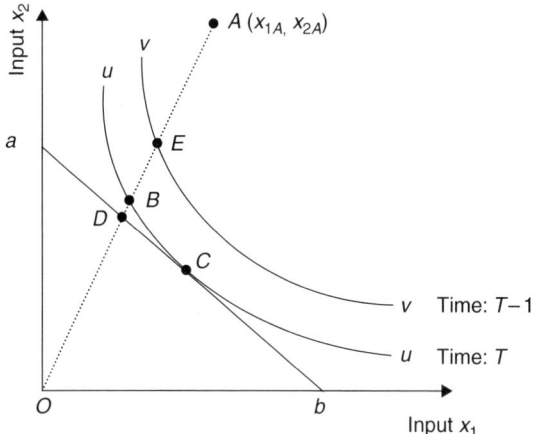

Figure 1.4 Comparison between technical and allocative efficiency.
Source: Drawn by the authors.

ratio *OB/OA*. Let *ab* represent the ratio of input prices. The ratio *OD/OB* then measures allocative efficiency, since the cost of point *D* is the same as that of the allocatively efficient point *C*, and is less than that of the technically efficient, but allocatively inefficient, point *B*.

All of the discussion above on efficiency and productivity is limited to a static perspective. By adopting a dynamic perspective, the set of production possibilities will exhibit a tendency to move over time, often due to the impact of technological innovation. When this implicit dynamic movement in the production possibility frontier occurs, it can be quite critical to determine whether or not a particular firm is keeping pace with such developments.

This basic idea can be illustrated in Figure 1.4: here, *vv* and *uu* stand for two production frontiers at two different points in time: *T* − 1 and *T*, respectively. The shift of the production frontier is caused by technological innovations that allow the same outputs to be produced with less input (*vv* → *uu*). Firms may or may not be able to adjust to these new production possibilities. For instance, one firm may improve its performance from point *E* at time *T* − 1 to point *B* at time *T*, implying that this firm has succeeded in immediately and properly incorporating the technical innovation into its production process. If another firm, however, takes no advantage of these new production technologies and its production process remains the same (i.e. held in steady state at point *E*), both before and after the industry-applicable technological innovation, then its performance as measured by technical efficiency will deteriorate relative to the production frontier at *T*.

Only technical efficiency is examined in depth in this book. Unless otherwise indicated, therefore, the term *efficiency* in this book should be taken to mean *technical efficiency*. Comparing the fundamental concepts and theoretical approaches to measuring efficiency and productivity in static and dynamic contexts, it is easy to conclude that measurements of productivity change and technical efficiency are related, especially when considering the dynamic context. The former measures the shift in the production possibilities, whereas the latter indicates the extent to which firms maintain their position relative to the position to which a shifting frontier has eventually moved. Grosskopf (1993) points out that this distinction is important because it reveals the relationship between productivity growth and technical change. This may be quite informative for policy makers; a slowdown in productivity growth, due to increased inefficiency or, alternatively, due to a lack of technical change, may prompt different directions in policy formulation. For instance, slow productivity growth due to inefficiency may be attributable to institutional barriers to the diffusion of knowledge on innovations. In which case, policies to remove these barriers may be more effective in improving productivity than policies directed at innovation *per se*.

1.3   Measuring performance in the container port industry

The challenges for contemporary container ports boil down to three aspects: the complicated nature of production at container ports, the stringent demands from port users and the complicated market structure of the port industry. As far as the first two aspects are concerned, with the increasing size and speed of container ships, the turnaround time for ships has become crucial for shipping lines in evaluating the attractiveness of a container port. In order to gain a competitive advantage and serve their main customers – the shipping lines – contemporary container ports are required to invest heavily in expensive, state-of-the-art equipment and to employ sophisticated management. Even so, container ports quite often find that they are at risk of losing important customers because of an adjustment in the shipping routes of liner companies and/or the exacerbated competition between ports, brought about mainly by the increasing extent of overlap between hinterlands that is due to the continued investment in, and consequent improvement of, inland logistics systems (Cullinane and Khanna, 2000). A number of arguments have been raised in recent years in relation to determining the most appropriate market structure for the port industry. This has involved a focus on the degree of competition, and the relationship of this to the level of private sector participation, in the container port industry. In addition, some governments have revised their port industry systems in recent years in an effort to optimise them. For instance, an increasing number of countries have privatised their port industry in order to achieve a more competitive market structure that, it is hoped and intended, will lead to improved efficiency and other benefits such as a reduced public sector funding requirement.

Given these challenges for the port industry, monitoring the performance of a container port (particularly relative to its immediate competitors) not only provides port operators with a powerful management tool, but also constitutes a most important input for informing port planning and operations at the regional and national level. Heaver (1995) points out that benchmarking performance is important for ports, irrespective of whether they face fierce or even little competition. When terminals serve largely local and well-defined hinterlands and face little competition, it is possible that a port authority can become captive to the actions, demands and approach of its extant terminal operators. Port authorities caught in this position could then become reluctant to take action that would be seen as posing a competitive threat to existing terminals. Conversely, in an environment in which external competition is extant or likely to arise in the future, greater onus is placed on the port authority to exert vigilance over, and to foster concern about, competitive performance. This might be achieved by a programme of benchmarking port and individual terminal performance. Such a programme would provide up-to-date information on technologies and the performance of systems in leading terminals around the world. The port authority might then see itself as a leader in 'dissatisfaction' and not as a defender of the *status quo*. In other words, port authorities should be proactive in regularly soliciting the views of its users and seeking out areas for improvement, rather than waiting to act in mere response to the complaints of shippers or shipping lines.

1.4 Problem statement

Despite the paramount importance of the container port industry in facilitating current and emergent trends in globalisation and international trade, there exists very little empirical work aimed at applying traditional economic theories to the container port industry and assessing their veracity. In particular, a study of traditional economic theory reveals that controversies exist in relationships between monopoly and competition and public ownership and privatisation, and their influence on a firm's performance. For instance, many economists believe that a competitive market structure encourages greater efficiency than alternative market structures (Leibenstein, 1973). On the other hand, it is argued that a competitive market structure is not necessarily sufficient for superior efficiency, as there may be other sources of efficiency, including economies of scale, that favour monopolistic industries (Schwartzman, 1973). These issues are important for the container port industry and should be subject to further investigation. To this end, after investigating the oligopolistic situation in Australia, Goss (1982) found that private sector ownership was associated with not only unjustifiably high port charges, but also that inefficiency and excessive costs were prevalent. However, Goss (1982) also found that both the ports of Hong Kong and of Singapore demonstrate high levels of efficiency, despite operating under very different market structures.

Given a proposition that traditional economic theory cannot provide the answer to all of the behavioural and performance aspects that relate to the container port industry, empirical proof in this aspect has become particularly important. From an empirical point of view, the performance of ports has been variously evaluated by calculating cargo-handling productivity at berth (Bendall and Stent, 1987; Tabernacle, 1995; Ashar, 1997), by measuring the productivity of a single factor (De Monie, 1987) or by comparing actual with optimum throughput over a specific period of time (Talley, 1998). It is clear that these traditional approaches are confined to partial measures of productivity and are insufficiently sophisticated to either reflect the complexity of contemporary container port production or to provide profound insights into the management or policy implications of container port (in)efficiency.

In recent years, two leading approaches to measuring efficiency have emerged, Data Envelopment Analysis (DEA) and Stochastic Frontier Analysis (SFA). Each of these techniques has occasionally been applied to the ports industry and, more specifically, to the container port industry. Compared with the traditional approach of focusing solely on partial productivity or even total factor productivity, efforts to derive a holistic measure of the efficiency of container ports constitutes a great leap forward in the in-depth analysis of container port production. However, extant research of this type is far from sufficient. *Inter alia*, the existing corpus of research is either based on strong assumptions or has commonly ignored the richness and diversity of the available data by utilising solely cross-sectional, as opposed to longitudinal data. The former may be biased because the strong assumptions may deviate from reality. Alternatively, in the case of the latter, it can be extremely misleading to ignore the role played by time. For instance, the impact of random effects or the effect of a recent investment in future production can both play a significant part in influencing the estimates of technical efficiency that are derived from the application of either DEA or SFA.

Apart from the two leading approaches, DEA and SFA, several other important approaches such as Free Disposal Hull (FDH) and various panel data approaches for measuring the efficiency of container ports have also been typically overlooked in the literature of port efficiency estimation. This is despite the fact that they are being increasingly applied to other industries and that they may produce useful insights.

1.5   Objectives

Against this background, this book aims to explore the economic theories underpinning container port production. Specifically, the relationship between ownership, competition and port efficiency will be examined by analysing the role of traditional economic theory, particularly in industrial organisation, to yield a rationale for the establishment of central government policies on port investment, policy and governance.

The various methodological approaches to estimating efficiency are outlined and evaluated as a precursor to justifying the derivation of more scientific and, therefore, convincing estimates of the efficiency of container port production. For the first time, comprehensive comparisons of alternative approaches to measuring efficiency will be undertaken for the container port industry. The techniques applied will include not only the two leading approaches, DEA and SFA, but also some of the other less well known but, nonetheless, important alternative approaches to measuring efficiency, such as the FDH method. The analysis contained in the book will also consider the outcomes of analyses based on panel data and taking into account random- and fixed-effects. Due to the individual strengths and weaknesses associated with the various alternative approaches to measuring efficiency, it is hoped and intended that this sort of comparative study represents a significant and timely contribution to the theoretical and empirical aspects of contemporary approaches to efficiency measurement.

1.6   Organisation of the book

This book is organised in the following manner. Chapter 2 focuses on examining fundamental and traditional economic theories underpinning container port production. Based on the theoretical arguments presented, a number of important hypotheses relating to the theoretical underpinnings of container port production are formulated and justified. These hypotheses constitute important constituents of the book that are tested at a later stage.

In Chapter 3, the available alternative approaches to measuring efficiency are reviewed, and their applicability to the container port context examined. This provides the methodological basis for a further empirical analysis of the technical efficiency of container ports.

Chapter 4 provides a definition of the input and output variables that are relevant to container port production and, in so doing, defines the empirical data requirements for the ensuing study. This is then followed by an explanation of the data collection process itself.

All of this work represents a necessary prequel for the actual empirical analysis that results in estimates of the efficiency of the container port industry as presented in Chapter 5. The implementation of alternative approaches is explained and illustrated and the robustness of the results derived from the various approaches is examined across the differing estimation techniques. In so doing, insights into the strengths and weaknesses associated with each of the approaches are deduced.

Chapter 6 mainly serves to compare empirical results and the inferences that can be drawn from these with the theoretical propositions and hypotheses that have been formulated and presented in Chapter 2. Assessing the explanatory power and ultimate validity of the body of economic theory underpinning container port production is undertaken with a view to determining the real-life relevance of conventional wisdom in theoretical economics.

Finally, Chapter 7 provides a holistic overview of the research that has been presented, draws conclusions and highlights potential future avenues for further research.

2
The Economic Theory of Container Port Production

2.1 Introduction

Understanding the economic theory that underpins container port production is a prerequisite for policy making or taking action in order to improve the performance of a container port. This is not an easy task, however, because of the complexity of container port production. This complexity is not only related to container port operations and their management, but also to the macroeconomic environment and market structure under which container ports operate. This chapter sets out to describe the economic functions of a port and the challenges for contemporary container ports arising from the complicated nature of their operations and the challenging market demands of port users and other market players.

This chapter continues by exploring the fundamental economic theory underpinning container port production; concepts that are derived from traditional industrial organisational theory. In this context, a differentiation is made between inter- and intra-port competition. This is because alternative forms of competition involve, and are affected by, a diversity of policies and lead to different managerial implications. Without such a differentiation, confusion is likely to arise.

This chapter contributes to the structure of this book by formulating some fundamental hypotheses concerning the economic theory underpinning container port production. These hypotheses constitute important suppositions that are examined in detail at a later stage in the book.

2.2   The economic functions of ports and their evolution

A port can be simply defined as a gateway through which goods and passengers are transferred between ships and shore (Goss, 1990a). However, contemporary ports and the functions they perform differ to a large extent from one port to another, or even from one time to another depending upon the particular stage in their evolution. It is difficult to provide a universal definition, therefore, that is applicable to all ports at different stages of their development. For instance, with the advent of global logistics, some European ports have extended their service provision to include inland transportation and third-party logistics (Heaver *et al.*, 2001; Notteboom and Winkelmans, 2001). At a minimum, however, a port should at least provide a suitable infrastructure, superstructure and service to its customers, as shown in Table 2.1, in order to fulfil the most basic of its functions.

Rather than remaining static, the port industry is constantly evolving over time. The evolution of the global ports sector is normally divided into three stages, as shown in Table 2.2. Up until 1960, a first-generation port constituted merely the cargo interface between land and sea transport. A second generation of ports emerged between the 1960s and the 1980s and involved their development into transport, industrial and commercial service centres. A third generation in port development emerged in the 1980s, principally due to a worldwide trend towards containerisation and greater intermodalism, combined with the growing requirements of international trade.

Ports have exhibited different production characteristics associated with the different stages of their evolution. For instance, at one stage in the past, port production was notoriously labour-intensive. However,

Table 2.1   Port facilities and services

Infrastructure	Approach channel, breakwater, locks and berths
Superstructure	Surfacing, storage (transit sheds, silos, warehouses), workshops, offices
Service to ships	Harbour Master's office (radio, VTS, etc.), navigational aids, pilotage, towage, berthing/unberthing, supplies, waste reception and disposal, security
Service to cargo	Handling, storage, delivery/reception, cargo processing, security

Source: UNCTAD (1995, p. 27).

Table 2.2 The development of ports

	First generation	*Second generation*	*Third generation*
Period of development	Before 1960	1960s to 1980s	After 1980s
Main cargo	Breakbulk	Breakbulk and bulk	Bulk and unitised
Attitude and strategy on port development	Conservative. Changing point of transport port	Expansionist Transport, industrial and commercial centre	Commercial Integrated transport node and logistics centre
Scope of activities	(1) ship/shore cargo interface	(1) + (2) cargo transformation Industrial activities	(1) + (2) + cargo and information distribution. Full logistics potential
Organisation characteristics	Independent activities, informal relationships	Closer relation between port and user. Loose relations in port activities. Casual relations between port and municipality	United and integrated relationships
Production characteristics	Cargo flow Low value added	Cargo flow and transformation Combined services Improved value added	Cargo/information flow and distribution Multiple service package. High value added
Decisive factors	Labour/capital	Capital	Knowledge/ know-how

Source: UNCTAD (1992, p. 23).

contemporary ports have dramatically improved their productivity, partially as a result of becoming highly capital-intensive. They now require, therefore, much less in terms of labour input (Suykens, 1983; Robinson and Reyes, 1998). Bearing in mind the dynamic nature of ports, the following hypothesis can be formulated.

Hypothesis 1: *The technical efficiency of container ports has improved with time.*

It is necessary at this point to emphasise that because of the complexity of the extensive activities carried out at container ports, the research reported in this book focuses solely on production at the level of the container terminal. The term 'port', therefore, refers to the aggregate activities of all terminals that operate within the area of the port.

2.3   The container port industry: the challenge of complexity

The operation of container ports can be considered one of the most complex tasks in the transportation industry. This is a consequence of (a) complex interactions (both physical and informational) between a number of different agents involved in importing and exporting containers (i.e. shipping companies, railroads and motor carriers, brokers, shippers, forwarders and regulatory agencies), and (b) complex operational interactions among the different service processes taking place at the port.

Container port operations can generally be divided into quay transfer operations along the berth, storage systems in container terminal yards, gate operations, and so on. As shipping lines are the most important clients of a port, the quay transfer operation fundamentally defines the efficiency of a port, and is vital to its competitive position. With the ever-increasing size of container ships (Cullinane and Khanna, 1999), it is not surprising to find a container ship with a carrying capacity of 6,000 TEUs (Twenty-foot Equivalent Units) calling at ports. Such large ships pose great challenges to transfer operations at the quay; a problem that is likely to increase as the average and maximum size of container ships continues to grow (McLellan, 1997).

As a storage area, the container yard acts as a buffer between sea and inland transportation or transshipment. The size of a ship is very frequently thousands of times the size of the land vehicles that carry the cargo to and from the port. As such, the use of storage is normally inevitable. Terminal operations are traditionally organised into three types of activities. The first involves the organisation of the stacking of the import containers and the unstacking of the export containers in order to achieve optimum movement to and from the container ship or quay and the container yard. The second activity is concerned with planning storage for all of the export containers that are expected to be received. The third activity involves responsibility over, and planning procedures for, the delivery of import containers (Chen, 1999).

Many containers in the yard have different destinations and properties, and are carried by different ships. To save on the amount of land utilised and to increase storage capacity, almost all container yards around the world stack their containers to several tiers. As such, the 'unproductive' movement of containers within a storage yard is a rather common, but undesirable, phenomenon. The operations and management strategies in the container yard ultimately influence the operational efficiency and operating cost of terminal operations as a whole. As a result, this aspect of container terminal operations has also proved to be a fertile area for research, as its interaction with what happens at the quay-side greatly influences the quality of the services offered and, therefore, the relationships formed with client shipping lines.

The major works investigating these aspects have, therefore, focused on such issues as:

- The optimal berth length and utilisation of quay cranes (e.g. De Weille and Ray, 1974; Noritake and Kimura, 1983, 1990; Schonfeld and Frank, 1984; Schonfeld *et al.*, 1985; Daganzo, 1989; Peterkofsky and Daganzo, 1990; Imai *et al.*, 1997; Musso *et al.*, 1999; Chu and Huang, 2002).
- Container ship turnaround times at port (e.g. Edmond and Maggs, 1976; Edmond and Maggs (1978); Robinson, 1978; Daganzo, 1990; Imai *et al.*, 1997; Imai *et al.*, 2001; Imai *et al.*, 2002a; Guan and Cheung, 2004).
- Approaches to optimum container ship stowage (e.g. Bischoff and Marriott, 1990; Avriel and Penn, 1993; Taleb-Ibraimi *et al.*, 1993; Chen *et al.*, 1995; Gehring and Bortfeldt, 1997; Avriel *et al.*, 2000; Wilson and Roach, 2000; Imai *et al.*, 2002b).
- Optimum yard operations (e.g. Chow, 1986; Lai and Lam, 1994; Evers and Koppers, 1996; Kim, 1997; Kozan, 1997; Kim and Bae, 1998; Kim and Kim, 1998, 1999).

Given the complexity of container port operations, there is a growing interest in applying advanced information technology to facilitate operations in container ports. In an advanced container port, before ocean-going vessels arrive, key information – container numbers, weights, size and type – is sent to the port via an electronic data interchange (EDI) and information exchange service. Ship planners then carefully plan vessel stowage for loading and discharge operations, using advanced technology and operating strategies. Yard planners decide the most appropriate yard areas for storing containers according to their designated vessel, weight class and port of discharge.

Heaver (1995) pointed out that economies of scale and of scope may exist in container ports through the use of sophisticated information technology because the value of this service is enhanced by the size and breadth of the network. De Neufville and Tsunokawa (1981) found empirical proof that a large container port is associated with high productivity and efficiency by studying the performance of the five major container ports of the east coast of the United States: Boston, New York–New Jersey, Philadelphia, Baltimore and Hampton Roads. Based on these assertions, a second hypothesis may be formulated as follows.

Hypothesis 2: *Efficiency increases as the scale of a container port increases. In other words, a large-scale container port is more likely to be associated with high efficiency than a small one.*

2.4   The challenge from other market players

Demand for port services is a derived demand and greatly influenced by other players in the market, mainly the port users. The port industry used to be perceived as a safe and non-competitive industry because past generations of ports exerted monopoly control over cargoes originating from or destined to their own hinterland. However, the rapid development of container transportation, the evolution of the global economic structure and the emergence of evermore sophisticated logistic systems have greatly changed this situation. Globalisation and supply chain management are playing an ever-increasing role in the contemporary global economy.

A basic driving force in this process is the structural shift in the economy from Fordism to post-Fordism. The Fordian era, named after Henry Ford, was based on the concept of 'economies of scale' in production through standardisation and the mass consumption of standard products. Since the 1980s, however, Fordism has gradually been replaced by a new kind of economic system that is directed more towards flexible organisations that cooperate in global economic networks. This shift in market structure causes producers to compete much more fiercely in the global market and is greatly reducing the possibilities for monopoly. The demand for non-standardised, better-quality goods with a short shelf-life means that supply chain management is extremely important to this process. This is because supply chain management is necessarily involved with issues of cost saving, value added logistics, the provision of efficient and satisfactory services, and so forth (Amin, 1994).

Table 2.3 Linkages between economic globalisation and the supply of transportation resources

Globalisation	Transportation
• Reliance on outsourcing, seeking labour and resource advantages • Flexibility in access to resources (regardless of distance) • Need for secure and efficient transaction • Just-in-time management of production and distribution processes • Zero inventory • Real-time information access and exchange • Opportunities for economies of scope • Customised production runs	• Longer and more customised transport linkages, often achieved through alliances across modes and across jurisdictions • Standardised equipment and procedures that allow quick and secure transactions • Great sensitivity to the timing of connections, arrivals and departures, and to the capacities of vehicles or carriage units • Speed of transactions, achieved largely through reductions in terminal transfer times • Expanded reliance on communications and computer networks for scheduling and tracking • Flexibility in modal choice and timing

Source: Revised from Janelle and Beuthe (2002).

This new market structure not only creates new opportunities for, but also exerts pressure over, the actors in the transport chain. This includes shipping lines, terminal operators, road hauliers and rail operators. In the post-Fordian era, therefore, transport services are becoming an integral part of production and marketing strategies. Since the demand for transport services is a derived demand, then all providers must follow logistics trends in order not to be left behind (Robinson, 2002). Table 2.3 singles out some of the most significant features of globalisation and attempts to match them with the attributes of modern transport systems. As explicitly emphasised by Janelle and Beuthe (2002), a related challenge exists for each of these linkages.

As a node within a transportation network system where many highly sophisticated logistics activities are concentrated, the modern container port is not only an interface between sea and inland transportation but, more importantly, should be able to facilitate freight flows throughout the whole network system. In fact, the specific characteristic requirements of container ports can be summarised as follows:

High productivity. This requirement is necessary from the perspective of a port's most important users, the carriers. Compared with traditional port production, containerisation has greatly improved the performance of port production. The contemporary container shipping industry is required to provide both economic and efficient services. In order to achieve economies of scale, numerous leading shipping lines around the world increasingly deploy large post-Panamax container ships. By consolidating and rationalising their fleets, shipping lines can optimise their routes using global hub-and-spoke transportation patterns. In so doing, the calls of large container ships and cargo distribution are concentrated in a few hub ports. A great deal of pressure is thus placed on these ports to handle cargoes efficiently (Hayuth, 1981; Cullinane and Khanna, 1999, 2000; Heaver *et al.*, 2000).

Low costs. Despite a rapid expansion in the size and number of shipping lines, the shipping industry has, for most of its history, been notoriously plagued by low profits as a result of overcapacity in the market and the fierce competition that exists between shipping lines (Brooks, 2000) and, in certain instances, with other modes of transport. A natural corollary for shipping lines has been to seek to negotiate lower priced services from the many container ports that often compete directly with each other for the transit of cargoes to and from a common hinterland.

Exposure to high risks. The knock-on effect of larger ships, fewer ports of call and the need for highly efficient facilities has forced previously non-competing ports into head-to-head competition as their hinterlands increasingly overlap (Cullinane and Khanna, 2000). Container ports, therefore, are continuously challenged, both physically and operationally. To maintain its competitiveness as a hub port, a container port has to invest heavily in sophisticated equipment or in dredging channels to accommodate the most advanced large container ships. Despite this, they are invariably at the mercy of shipping lines and risk losing important customers for various reasons. For instance, the merger of two lines may lead to an adjustment of routes, where some port calls will be abandoned. Under such circumstances, the only rationale that a port has for investment is best described in Slack (1993):

> This is somewhat analogous to a lottery, where only those who purchase tickets have a chance of winning ... . Even the largest ports have become pawns rather than the dominant players in the world-wide transportation game.

The high-risk nature of the container port industry may cause some port facilities to become redundant because of overcapacity or because of loss of important customers. This leads to the formulation of the following:

Hypothesis 3: *Fluctuations in the technical efficiency of container ports are not related to scale of production.*

2.5 Traditional theories of industrial market structure

According to Scherer and Ross (1990), a traditional topic in the field of economics is 'industrial organisation' or its equivalent 'industrial economics' (the latter term being more frequently used in Europe). Industrial organisation is fundamentally concerned with the relationship between productive activities and the organising mechanism. Among others, 'competition' and 'monopoly' are the two most important concepts in the context of industrial organisation.

Two controversies exist as to the relationship between market structure and efficiency. On the one hand, Schwartzman (1973) has argued that a competitive market structure is not necessarily sufficient to ensure superior efficiency, as there may be other sources of efficiency including scale economies that favour monopolistic industries. On the other hand, many economists believe that a competitive market structure encourages greater efficiency than alternative market structures (Leibenstein, 1973). Schwartzman's view has been strongly challenged by Leibenstein (1973):

Despite the fact that the conventional assumptions of microtheory lead to the conclusion that a monopolistic firm and a competitive firm are equally X-efficient, I would imagine that relatively few economists seriously believe this to be true.

Scherer and Ross (1990) pointed out that studying problems of industrial organisation has a direct and continuing influence on the formulation and implementation of public policies. These policies include the choice between private and public enterprises, the regulation or deregulation of public utilities, the promotion of competition through antitrust and free trade policies, and the stimulation of technological progress through the granting of patents and subsidies, and so on.

2.6   Port competition

The work of Scherer and Ross (1990) provides valuable guidance for any discussion of the container port industry and its market structure. The market structure of the container port industry can be analysed from the point of view of a nation (or continent, or even from a global perspective) or as an individual port. The former refers to a situation in which a port is regarded as a unit under a national (or even higher) level administration and competes or cooperates with other ports, while the latter refers to the various parties and their relationship within a port. As competition is one of the most important concepts in the context of market structure, the following two subsections are dedicated to differentiating between types of container port competition and the policy and management implications that are associated with each sort of competition.

2.6.1   Inter-port competition

2.6.1.1   Definition

Inter-port competition can be simply understood as the competition among (or between) different ports (within the context of this work, the discussion is obviously limited solely to container ports). The most important criterion for judging whether two container ports are competing with each other is to examine whether they serve the same or overlapping hinterland. From this point of view, hackneyed statements that relate to the 'competition between the ports of Hong Kong and Singapore' (as is often implied by the industry's trade press, as well as in certain academic works; e.g. Fung, 2001) should not be taken too seriously, since these two ports serve the trade of vastly different hinterlands; while the gateway port of Hong Kong serves mainly the cargo flows that originate from or are destined for southern China, the load centre port of Singapore mainly serves the cargo flows to and from Southeast Asian countries, such as Malaysia and Indonesia, as well as the North–South trades to Australasia.

Goss (1990d) points out that the extent to which a port is subject to competition varies according to a number of factors, including its geographic location and the nature of the goods that move through it. Competition is correspondingly restricted whenever there are few ports to compete with one another or when there are few ports handling a given commodity. Traditionally, inter-port competition was regarded as minor before the era of containerisation. Port markets used to be perceived as being monopolistic or oligopolistic, because of the exclusive and immovable geographical location of the port and the unavoidable concentration of port traffic that this brought about. The rapid development of international containerised and intermodal transportation has, however, drastically changed this situation. Many container ports no longer enjoy the freedom yielded by a monopoly or oligopoly over the handling of cargoes from within what might be considered to be their hinterland. They are not only concerned with whether they possess the physical capacity to handle the cargo, but also whether they are able to successfully compete for it (Notteboom *et al.*, 1997; Coeck *et al.*, 1997; Winkelmans, 1998, 2002; Huybrechts *et al.*, 2002).

A poignant example of this, provided by Talley (2000), is the inter-port competition in the United States, where container ports are now finding themselves competing more intensely, not only against nearby rivals, but also against ports hundreds of miles away. It used to be unimaginable that ports in the east and west coasts of the United States could compete with each other. However, this has been changed by the emergence of landbridge services. Until the early 1980s, most ocean container cargo sourced from Asia and bound for the eastern United States was shipped across the Pacific, through the Panama Canal, to a port on the east coast of the United States. In April 1984, however, the routing of this cargo began to change. Ships began calling at ports along the west coast of the United States, where containers were unloaded and put on rail cars for transit towards the east.

Earlier work on port competition mainly consists of the studies of Verhoeff (1981) and Goss (1990d). Both of these studies focused, however, on the port industry as a whole, rather than on container ports in particular. Moreover, the container port industry has changed dramatically in the past few years. Nevertheless, the work of Verhoeff and Goss provides a useful framework for a discussion of inter-port competition. A refinement of their work with reference to the contemporary container port industry yields the following major forms of inter-port competition, listed in decreasing order of the size of the geographic range of the competitive ports.

Competition between whole ranges of ports or coastlines. A typical example of contemporary inter-container port competition is that between ports on the west coast of North America that are in competition with ports on the east coast. This competition has been greatly enhanced by the development of both intermodal systems of transfer and more efficient forms of long-distance transport. Another example may be provided by the increasing evidence (and national strategic ambitions of the nations concerned) that competition between the ports in the Hamburg–Le Havre range and those of the Southern Mediterranean is becoming more intense as the latter strive to deprive the former of Asia sourced and bound cargoes that have destinations or origins in inland Europe (Alberghini, 2002).

Competition between ports in different countries. This kind of competition is prominent in several European countries, such as along the Hamburg–Le Havre range and most especially between Antwerp in Belgium and Rotterdam in the Netherlands. Another example is provided by the container ports of the Pacific North-West where intense competition exists between the container port of Vancouver in Canada and those of Tacoma and Seattle in the United States.

Competition between individual ports in the same country. This kind of competition occurs frequently when several ports are similar and their hinterlands overlap. Examples include the competition in the United States between the port of Oakland and the nearby port of San Francisco, and between Long Beach and Los Angeles in California. In Northern China, similar competition exists between the container ports of Dalian, Qingdao and Tianjin in the Bay of Bohai. Also in China, there is increasing competition between the ports of Shanghai and Ningbo on the Eastern-Central seaboard, as the latter eats into the market share of the former (Cullinane, 2003a, 2004).

2.6.1.2   *Discussion*

As in many other industries, debates also exist in the container port industry as to the relationship between inter-port competition and port performance. Advocates of competition believe that it can encourage innovation, increase the sense of responsibility of staff, free an organisation from the constraints of bureaucracy, and promote high efficiency. It seems that policies to encourage inter-port competition are gradually being accepted by an increasing number of governments. Heaver (1995) reports that the policies of governments are moving in ways consistent with a more competitive market structure brought about by decentralisation.

However, many economists and governments also appreciate the advantages of a monopolistic market in the (container) port industry, equivalently brought about by a policy of centralisation. Among others, Heaver (1995) has argued that the main advantage of a central planning policy for the port industry is to avoid the risk of excess capacity. The capital intensity and long life of container terminals, the interest of ship owners to minimise ship delays and the optimism of terminal and port managers are suggested as reasons why container ports may exhibit overcapacity. Although this applies to ports falling under the control of a central government, local authority or similar public regime, the risk of surplus capacity may be aggravated when local authorities are responsible for ports. For example, in relation to port privatisation, Turnbull and Weston (1993, p. 119) suggested that policy changes in the United Kingdom have 'not been sufficient to resolve a number of the industry's more persistent, underlying problems. ... Most notable among these are over-capacity, the duplication of the investment, and the zero sum or redistributive nature of competition.'

Another drawback of inter-port competition is that it may cause a port to assume high risks. To maintain its competitiveness in a competitive market, a port has to invest heavily in sophisticated equipment or in dredging channels to accommodate the most advanced and largest container ships. Despite this, in a competitive environment where shipping lines have the choice between more than one port to use, a port faces the risk of losing its important customers. Some economists, however, challenge whether this is really a drawback associated with the container port industry (e.g. Heaver, 1995), because the port industry is not unique in the lumpiness of its investments. A similar situation arises in many other industries. As such, this line of argument suggests that the container port industry has no more reason to protest against the costs of forecasting errors and lumpiness in capacity than any other industry.

2.6.1.3   Policy and managerial implications

The relevant policies on inter-port competition should be analysed taking into account the levels of competition suggested in Section 2.6.1.1. If inter-port competition takes place across several countries, it might be difficult to use policies or regulations to encourage or discourage competition. This, however, does not necessarily imply that governments can do nothing to influence inter-port competition. Along with others, Heaver (1995) has suggested that the harmonisation of international policies is particularly important for the contemporary container port industry because 'the distorting effects of differential subsidies are much more likely to extend across national boundaries as the size of the hinterland is increased'.

The managerial implications in this respect include performance benchmarking. Such an activity enables competing ports to obtain up-to-date information on technologies and on the performance of systems in leading container ports around the world, to ensure that they are not lagging behind. Another important managerial implication is that container ports should plan and invest in capacity in a sophisticated manner, taking into account the lumpiness of capacity and the long life of major infrastructure and facilities. In the dynamic market that characterises the contemporary container port industry, however, this is easier said than done.

Within a reasonable arena such as exists within national boundaries or perhaps within economic blocs, it is possible that governments can devise policies or regulations to guide inter-port competition. An interesting question that then arises is whether inter-port competition should actually be encouraged or, alternatively, whether competitive ports would be better off cooperating with other ports?

National policies governing inter-port competition largely arise from, and are dependent upon, whether the port is under the direct control of either central or local government and usually correspond to policies of centralisation and decentralisation, respectively. If the whole industry is controlled by the central government, it will normally formulate policies to benefit the overall port or transportation network, sometimes at the cost of individual ports within the national sector. If, on the other hand, the port is locally governed, the objective of the port will quite likely involve pursuing a combination of its own and local/regional interests. It is clear that, in the latter situation and under the right circumstances, inter-port competition will intensify.

When competitive ports exist within the same country, the central government can exert a greater influence on these ports. It is clear, however, that the majority of governments actively promote some degree of inter-port competition. This is reflected in the fact that, in most developed countries, ports have been given an increased degree of independence and autonomy based on the belief that this system enables the ports to be more economic, more innovative, to exhibit greater responsibility towards their customers and to adjust themselves quickly in a dynamic market. Most central governments will still, however, closely monitor the performance of ports and retain the final say in important decisions such as adding new capacity or the building of new ports. In so doing, it is hoped that economies of scale or scope can be achieved, and the duplication of construction and waste can be avoided.

Within the same country, it is more likely that two or more competitive ports can cooperate with each other. Cooperative arrangements may vary from joint marketing and investment to even closer forms of cooperation, ultimately culminating in merger and acquisition. Song (2002, 2003) argued that, in order to attain a 'win–win' outcome, the container port industry should attach more importance to *co-opetition* – a combination of competition and cooperation that was conceptualised in the works of Jorde and Teece (1989) and Brandenburger and Nalebuff (1996). Similar suggestions can be found in Heaver *et al.* (2001).

Based on the above discussion, a further hypothesis focuses on the relationship between inter-port competition and port performance. It should be emphasised that this hypothesis is rather a tentative one because, as mentioned previously, traditional economic theories lead to different conclusions about inter-port competition and efficiency.

Hypothesis 4: *Container ports faced with external competition are more efficient than their counterparts in a less competitive market.*

2.6.2   Intra-port competition

Goss (1990c, p. 256) provided an explicit explanation of intra-port competition:

> this [intra-port competition] does not necessarily mean that there should actually be a large number of firms competing simultaneously: it means that the markets in question should be contestable, in the sense that entry is easy for a new firm, whose exit will also be easy if its efforts turn out to be unsuccessful.

2.6.2.1   Ownership and administration

Discussions of intra-port competition are necessarily related to port ownership and administration because there is essentially no competition within a single port if the whole port is under the control of a public authority. Ports can be classified according to their type of ownership or administration. The allocation to a variety of parties of property rights over the infrastructure, superstructure and services gives rise to different patterns of port ownership. Goss (1990b) was probably the first to attempt to address this issue. According to the role played by the port authority, ports are divided into three types: (a) a 'comprehensive port' if the port authority itself performs all, or almost all, of the activities carried on within the area of the port; (b) a 'landlord port', if the port authority is only responsible for planning the port and exercising overall control over the activities carried on within it, but delegating these extensively to private sector companies. Goss (1990b) argues that the landlord port is the opposite of the comprehensive port; (c) a 'hybrid' port lies somewhere in between.

This classification of ports was further developed by Liu (1992). According to Liu (1992), a port can be categorised as: (a) a 'service port' if the port authority is responsible for the provision of all port facilities and services; (b) a 'tool port' if the public port authority provides the infrastructure and superstructure, while the provision of services is licensed to private operators; (c) a 'landlord port', in which the domain of the port authority is restricted to the provision of the infrastructure, while investment in the superstructure and port operation is the responsibility of licensed private companies; and (d) a 'private port' if the provision of all of the facilities and services is left to the private sector.

Another alternative for analysing port administration and ownership was proposed by Baird (1995, 1997) and refers to a port function matrix as shown in Table 2.4.

The starting point for this conceptual framework is that a port must fulfil and provide the following three functions, whether it is in private or public hands:

- *A regulatory function*, which can involve the granting of substantial powers to the port's public or private sector management, the majority of which will be of a statutory nature. In general, this function may be regarded as the primary role of a port authority.
- *A landowner function*, expedited by allowing ports to control significant areas of land. Irrespective of whether the land area of a port is large or small, however, the essential tasks involved would be to manage and develop the port estate; to implement policies and strategies for the port's physical development in terms of superstructure and (sometimes) infrastructure; to supervise major civil engineering works; to coordinate port marketing and promotion activities; to provide and maintain channels, fairways, breakwaters, and so on; to provide and maintain locks, turning basins, berths, piers and wharves; and to provide or arrange for road and rail access to the port facilities.
- *An operator function*, which is concerned with the physical transfer of goods and passengers between sea and land.

Table 2.4   Port function matrix

Port models	Port functions		
	Regulator	*Landowner*	*Operator*
PUBLIC	Public	Public	Public
PUBLIC/private	Public	Public	Private
PRIVATE/public	Public	Private	Private
PRIVATE	Private	Private	Private

Source: Baird (1995, 1997).

According to which of these three functions are the responsibility of public or private organisations, the matrix presented in Table 2.4 makes it possible to ascertain the extent of the influence exerted by the public and private sectors within any given port. The matrix also suggests the four main patterns (as defined by choice of port administration – ownership, management and operation) into which a government is able to organise its port industry.

According to Table 2.4, port administration and ownership models are divided into four types of port administration: the PUBLIC port, the PUBLIC/private port with the public sector dominant, the PRIVATE/ public port with the private sector dominant, and the PRIVATE port. In a PUBLIC port all three functions are controlled by the government or public authority. In the PUBLIC/private port, the operator function is controlled by the private sector, with both the regulatory and landowner functions remaining in the hands of the government. In the PRIVATE/ public port, both the landowner and operator functions are in private hands, while the regulatory function remains within the public sector. Finally, in the PRIVATE port, all three essential functions are controlled by the private sector.

Although traditional industrial organisation theory states that competition has its pros and cons in any organisation, unlike the debates over whether inter-port competition should be encouraged or discouraged, there is almost universal agreement that intra-port competition leads to improved port performance, and therefore should be encouraged. Goss (1990c) suggests that 'comprehensive' ports achieve their efficiency by direct management. Landlord ports, meanwhile, achieve efficiency by introducing competition, and the role of a port authority is to guarantee the existence of competition (Goss, 1990c). Similarly, Heaver (1995) has argued that

> encouraging competition among terminals even in the same port, rather than discouraging competition appears to be the appropriate policy direction.

Among others, privatisation is a useful approach to introducing intra-port competition (Song *et al.*, 2001). The global trend towards the privatisation of ports in order to improve their economic efficiency perhaps reflects the increasing awareness of the importance of intra-port competition in achieving this. However, privatisation cannot always guarantee improved port performance, as illustrated in Figure 2.1.

Figure 2.1 shows that because of the lumpiness of port and terminal capacities and the long lifespan of major port facilities, port privatisation is normally accompanied by a long contract between private enterprises and port authorities or government. As such, a new monopoly or oligopoly within the port may develop. If there is neither inter- nor intra-port competition (or even simply the threat of it), it is difficult to conclude whether public management will outperform private management, or the other way around. Some controversy exists as to the relationship between port ownership and performance under this circumstance. Some economic theorists argue that public management will do better in terms of economic efficiency than private management (e.g. Vickers and Yarrow, 1988). Goss (1982, 1990a–d) provides empirical evidence supporting the theory, however, that in oligopolistic (or monopolistic) markets, private sector port operators are very likely to lose the incentive to improve performance. Analysing the situation of oligopoly in Australia, he found that private sector ownership was associated not only with unjustifiably high port charges, but also that inefficiencies and excessive costs were prevalent. The similar failure of port privatisation in the United Kingdom has also been reported (Saundry and Turnbull, 1997).

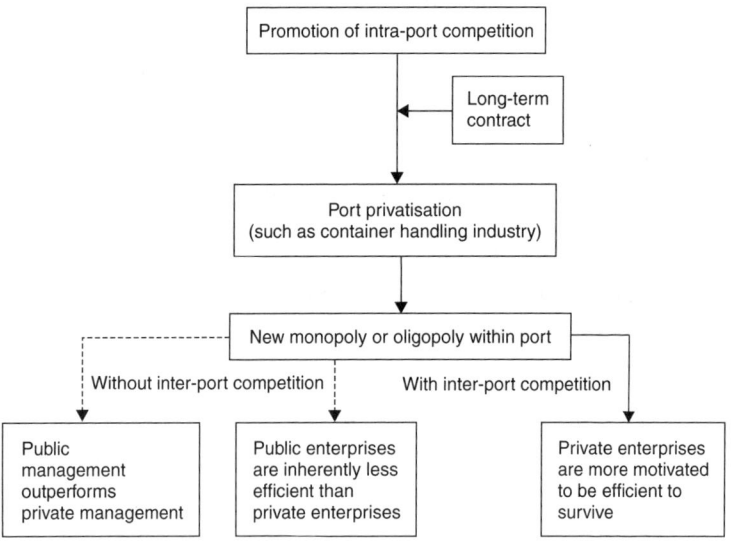

Figure 2.1   Privatisation and port performance.

Source: Drawn by the authors.

Hence, if this market structure prevails, public ports should perform better than their private sector counterparts.

On the other hand, public ports may suffer deficiencies as a result of the public monitoring hierarchies that hold sway over them, goal displacement, lack of clarity in corporate objectives and operative responsibility, and excessive ministerial intervention in operational decisions (Liu, 1992, p. 65). From this point of view, economists argue that public enterprises are inherently less efficient than private enterprises, especially where the market is competitive. This is often purported to be the case in relation to container ports and/or terminals (see Song *et al.*, 2001).

Figure 2.1 also shows that given a situation of inter-port competition, which is particularly characteristic of the environment in which modern container ports now operate, private enterprises are more likely to outperform their public counterparts. This is primarily the case because in seeking to maximise their own benefit, private enterprises are more motivated to seek viable ways to survive and prosper in a competitive market.

Based on this discussion, another tentative hypothesis referring to port ownership and efficiency can be formulated:

Hypothesis 5: *The efficiency of container ports improves as ownership moves towards greater private sector participation.*

2.6.2.2   Policy and managerial implications

As agreed by many experts, ports are significant not only because of their contribution to economic welfare, but also because of their military significance. As such, port privatisation cannot be treated in the same way as other industries. In most cases, national interests are a major concern.

National port policy should aim to improve the performance of the whole port industry within that country. Intra-port competition takes place within a port, and is therefore not greatly influenced by certain aspects of national policy. Port authorities should ensure, however, that the internal market within the port is contestable in order to encourage improved performance. In addition, a port authority should play a leadership role in facilitating cooperative activities to achieve economies of scale or scope within the whole port and its environs, a role that is sometimes not within the capability of a single firm within the port.

Performance benchmarking is also a very important tool for helping to guarantee port performance. This activity benefits both port authorities and firms (whether they be public or private) involved in port production.

Through appropriate and well-considered benchmarking, port authorities are able to assess whether their production is efficient compared with that of their counterparts, or whether one entity is more efficient than another within the port. At the same time, companies involved in port production are able to ascertain how to potentially improve their production and enhance their competitiveness. Based on performance benchmarking theory, another hypothesis can be proposed as follows:

> **Hypothesis 6:** *Different firms within a port have the same or similar efficiency.*

2.7   Summary

In this chapter, traditional industrial organisation theory is applied to the analysis of container port production. In particular, the two fundamental paradigms of 'competition' and 'monopoly' that lie at the core of traditional organisation theory are examined with reference to the container port industry.

Port competition plays an important role in the contemporary container port industry. In this chapter, port competition is classified as inter- and intra-port competition. This dichotomous classification underpins a further discussion of the economic theory that is relevant to container port production. Such classification is significant because it not only avoids possible confusion but, more importantly, can yield in-depth insights into the container port industry.

Based on this theoretical discussion of the economic theory underpinning container port production and the ensuing policy and managerial implications, this chapter formulates six hypotheses. These hypotheses constitute important theoretical arguments for probing the nature of container port production and are formally tested at a later stage in this book.

3
Alternative Approaches to Efficiency Measurement

3.1 Background

This chapter aims to investigate the various approaches to efficiency measurement and to lay the required methodological groundwork for analysing the productive efficiency of a sample of container ports. The relative advantages and weaknesses associated with each of the available alternative approaches, as well as their applicability to this particular context, are critically reviewed. In order to determine the potential for instigating methodological improvements, previous applications of these alternative approaches to the port industry, particularly the container port sector, are also examined.

As discussed in Chapter 1, a firm's efficiency is easily measured by comparing its productivity with that of a firm that produces on the production or cost frontier. Under this concept, those firms that operate on either a (maximum) production frontier or a (minimum) cost frontier are deemed to be 'efficient'. In contrast, inefficient firms operate either below the industry production frontier or above the industry cost frontier. Relative to a firm located on a production frontier, an inefficient operator will produce less output for the same cost. Analogously, relative to any firm located on a cost frontier, an inefficient operator will produce the same output for greater cost.

A production frontier is normally unknown and, in practice, can only be *estimated*. The literature on frontier models was inaugurated in the seminal contribution of Farrell (1957), who provided a rigorous and comprehensive framework for analysing efficiency across firms. This was achieved in terms of evaluating realised deviations from an idealised frontier isoquant that was derived from a given sample, with the methodology introducing an approach based on developing a piecewise linear envelopment of the data in order to determine the specification of the frontier. In summary, Farrell's efficiency measurement is based on the following three principles:

- efficiency measurement is based on radial uniform contractions or expansions from inefficient observations onto the frontier;
- the production frontier is specified as the most pessimistic piecewise linear envelopment of the data;
- the frontier is calculated through solving systems of linear equations, obeying the two conditions on the unit isoquant:
  (a) That the slope of the frontier is not positive;
  (b) That no observed point lies between the frontier and the origin.

Figure 3.1 provides a conceptual illustration of Farrell's approach where two inputs x_1 and x_2 are used to produce the same amount of output. The piecewise linear frontier in Figure 3.1 is defined by *ff* and the efficiency of firm *A* is measured by *OB/OA*.

Since Farrell (1957) introduced his seminal work, it has been followed by a relatively large number of refinements and extensions (Forsund *et al.*, 1980; Forsund and Sarafoglou, 2002). Figure 3.2 summarises the alternative approaches to measuring efficiency that are based on the work of Farrell (1957) and which utilise purely cross-sectional data. Clearly, over decades of development, a large number of more sophisticated approaches to frontier efficiency measurement have been proposed.

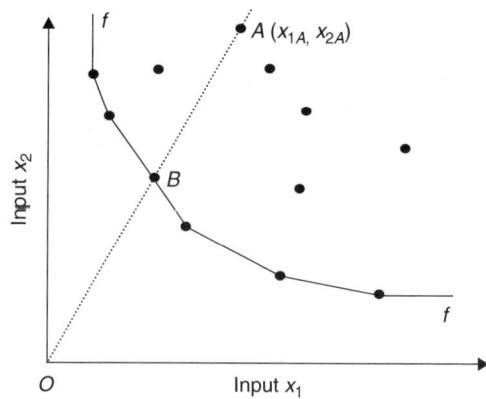

Figure 3.1   Farrell's measurement of efficiency.
Source: Drawn by the authors.

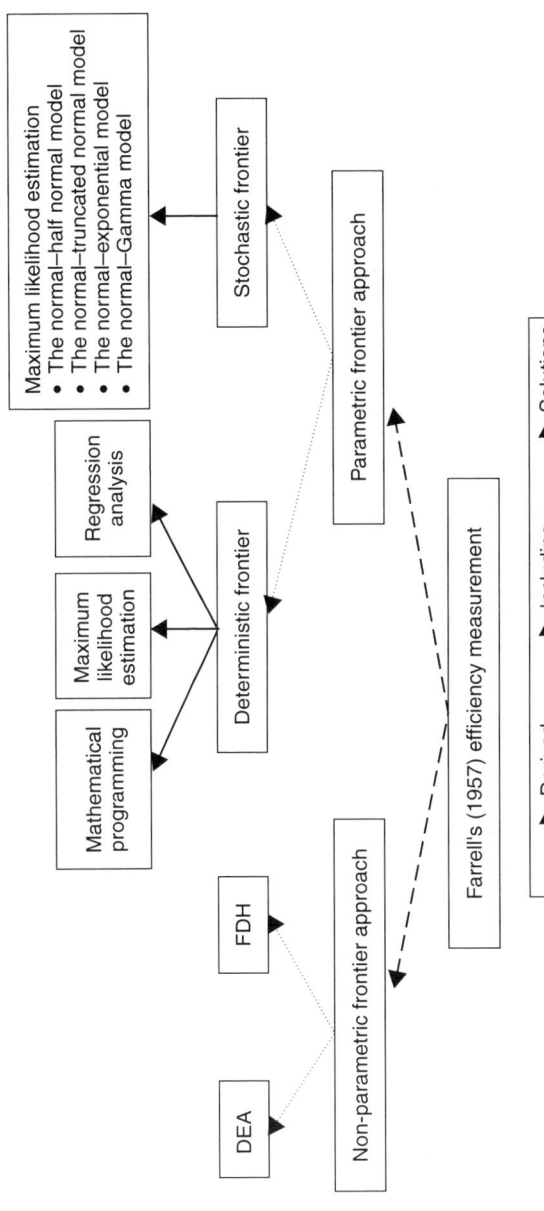

Figure 3.2 Alternative approaches to efficiency measurement derived from Farrell (1957).

Source: Summarised by the authors.

Among other bases, the way in which the production (or cost) frontier is constructed and the nature of the data used in the model are just two of the possible criteria that may be applied for classifying the alternative approaches.

In terms of the way in which the frontier is constructed or estimated, the frontier approach mainly utilises econometric and mathematical programming techniques. The differences between these broad estimation procedures can be summarised as follows (De Borger *et al.*, 2002):

- *Parametric (econometric) versus non-parametric (mathematical programming) frontier specifications*: The parametric approach assumes that the boundary of the production possibility set can be represented by a particular functional form with constant parameters, while the non-parametric approach imposes minimal regularity axioms on the production possibility set and directly imposes (or constructs) a piecewise technology on the sample.
- *Stochastic (econometric) versus deterministic (mathematical programming) frontier specifications*: Stochastic methods make explicit assumptions with respect to the stochastic nature of the data, by allowing for random measurement error. Deterministic methods, on the other hand, take all observations as given and implicitly assume that these observations are exactly measured.

Many studies combine the above distinctions (parametric versus non-parametric, stochastic versus deterministic) to yield a four-way classification, as illustrated in Table 3.1.

Table 3.1 Taxonomy of frontier methodologies

Functional form	Measurement error	
	Deterministic	Stochastic
Parametric	Corrected OLS, etc.	Frontiers with explicit assumptions (exponential, half-normal, etc.) for the technical efficiency distributions
Non-parametric	FDH, DEA-type models, etc.	Resampling, chance constrained programming, etc.

Source: De Borger *et al.* (2002).

As far as the data are concerned, for the past several decades, most efficient measurement approaches mainly utilise cross-sectional data. In recent years, however, panel data have been increasingly utilised because of the numerous advantages that panel data possesses over cross-sectional data. This is despite the computational complexity involved in estimating frontiers on the basis of panel data observations. As stated by Kumbhakar and Lovell (2000, p. 10), 'Cross-sectional data provide a snapshot of producers and their efficiency. Panel data provide more reliable evidence on their performance, because they enable us to track the performance of each producer through a sequence of time periods.'

This chapter is structured as follows. In Section 3.2, the notation and symbols used in this book are introduced. Sections 3.3 and 3.4 discuss two non-parametric approaches; DEA and FDH. In each section, DEA and FDH are discussed in relation to whether cross-sectional or panel data are used in the model. Sections 3.5 and 3.6 focus on another stream of efficiency measurement: the parametric frontier approach, and are organised in relation to not only the nature of the data (i.e. whether the data are cross-sectional or panel), but also in accordance with the different assumptions imposed on the efficiency estimates and the estimation errors that are derived from applying the different approaches to efficiency estimation. In Section 3.7, previous applications of the various approaches to the port or container port industry are examined. In so doing, the deficiencies in these previous applications are identified and further research opportunities outlined. Finally, Section 3.8 summarises this chapter.

3.2   Notation and symbols used

In order to avoid any possible confusion caused by the different usage of notation, throughout this book the same symbol is consistently used to denote the same variable or parameter. For instance, x and y denote, respectively, the input and output variable in alternative models. The use of matrices is inevitable in some complicated computations. The most important matrices used in this book are presented in equations (3.1) and (3.2) which, in the context of cross-sectional and panel data, respectively denote situations where there are a number of firms using M inputs to produce N outputs. In essence, equation (3.2) is a particular form of equation (3.1) when the study period is $T = 1$. It is still presented separately, however, so as to simplify the discussion of the situation when cross-sectional data are used. In equation (3.2) x_{mk} refers to the mth input data of firm k, whereas y_{nk} is the nth output of firm k. In equation (3.1) x_{mkt} and y_{nkt} refer to the mth input and the nth output of firm k at time t.

When an approach based on the parametric econometric model is used, normally only one output variable is considered. In this case, the output variable **Y** is rewritten as shown in equations (3.3) to correspond, respectively, to the cross-sectional and panel data cases. Equations (3.4) and (3.5) show, respectively, the logarithmic values of a one-sided efficiency component **u** and a two-sided random effects component **v** in econometric models, under the respective assumptions in each case that efficiency is time-invariant and time varying. In the case of time-invariant efficiency, the number of values of both u_k and v_k is T. The superscript ['] in equations (3.4)–(3.5), and hereafter, refers to the transpose of the matrix. Note that $U = \exp(-u)$ and $V = \exp(-v)$ denoting, respectively, the real efficiency and random effects.

$$X = \begin{bmatrix} x_{111} & x_{211} & \cdots x_{m11} & \cdots x_{M11} \\ x_{112} & x_{212} & \cdots x_{m12} & \cdots x_{M12} \\ \vdots & \vdots & \vdots & \vdots \\ x_{11T} & x_{21T} & \cdots x_{m1T} & \cdots x_{M1T} \\ \vdots & \vdots & \vdots & \vdots \\ x_{1k1} & x_{2k1} & \cdots x_{mk1} & \cdots x_{Mk1} \\ \vdots & \vdots & \vdots & \vdots \\ x_{1kt} & x_{2kt} & \cdots x_{mkt} & \cdots x_{Mkt} \\ \vdots & \vdots & \vdots & \vdots \\ x_{1kT} & x_{2kT} & \cdots x_{mkT} & \cdots x_{MkT} \\ \vdots & \vdots & \vdots & \vdots \\ x_{1K1} & x_{2K1} & \cdots x_{mK1} & \cdots x_{MK1} \\ \vdots & \vdots & \vdots & \vdots \\ x_{1KT} & x_{2KT} & \cdots x_{mKT} & \cdots x_{MKT} \end{bmatrix}_{(K \times T) \times M}' \qquad Y = \begin{bmatrix} y_{111} & y_{211} & \cdots y_{n11} & \cdots y_{N11} \\ y_{112} & y_{212} & \cdots y_{n12} & \cdots y_{N12} \\ \vdots & \vdots & \vdots & \vdots \\ y_{11T} & y_{21T} & \cdots y_{n1T} & \cdots y_{N1T} \\ \vdots & \vdots & \vdots & \vdots \\ y_{1k1} & y_{2k1} & \cdots y_{nk1} & \cdots y_{Nk1} \\ \vdots & \vdots & \vdots & \vdots \\ y_{1kt} & y_{2kt} & \cdots y_{nkt} & \cdots y_{Nkt} \\ \vdots & \vdots & \vdots & \vdots \\ y_{1kT} & y_{2kT} & \cdots y_{nkT} & \cdots y_{NkT} \\ \vdots & \vdots & \vdots & \vdots \\ y_{1K1} & y_{2K1} & \cdots y_{nK1} & \cdots y_{NK1} \\ \vdots & \vdots & \vdots & \vdots \\ y_{1KT} & y_{2KT} & \cdots y_{nKT} & \cdots y_{NKT} \end{bmatrix}_{(K \times T) \times N}$$

$$(3.1)$$

$$X = \begin{bmatrix} x_{11} & x_{21} & \cdots x_{m1} & \cdots x_{M1} \\ x_{12} & x_{22} & \cdots x_{m2} & \cdots x_{M2} \\ \vdots & \vdots & \vdots & \vdots \\ x_{1k} & x_{2k} & \cdots x_{mk} & \cdots x_{Mk} \\ \vdots & \vdots & \vdots & \vdots \\ x_{1K} & x_{2K} & \cdots x_{mK} & \cdots x_{MK} \end{bmatrix}_{K \times M}' \qquad Y = \begin{bmatrix} y_{11} & y_{21} & \cdots y_{n1} & \cdots y_{N1} \\ y_{12} & y_{22} & \cdots y_{n2} & \cdots y_{N2} \\ \vdots & \vdots & \vdots & \vdots \\ y_{1k} & y_{2k} & \cdots y_{nk} & \cdots y_{Nk} \\ \vdots & \vdots & \vdots & \vdots \\ y_{1K} & y_{2K} & \cdots y_{nK} & \cdots y_{NK} \end{bmatrix}_{K \times N}$$

$$(3.2)$$

$$\mathbf{Y}' = \begin{bmatrix} y_1 \ y_2 \cdots y_k \cdots y_K \end{bmatrix}_{1 \times K} \quad \text{(cross-sectional data)}$$

(3.3)

$$\mathbf{Y}' = \begin{bmatrix} y_{11} y_{12} \ \cdots \ y_{1T} \ \cdots \ y_{k1} \ \cdots \ y_{kt} \ \cdots \ y_{kT} \ \cdots \ y_{K1} \ \cdots \ y_{KT} \end{bmatrix}_{1 \times (K \times T)}$$
(Panel data)

$$\mathbf{u}' = \begin{bmatrix} u_{11} \ u_{12} \cdots u_{1T} \cdots u_{k1} \cdots u_{kt} \cdots u_{kT} \cdots u_{K1} \cdots u_{KT} \end{bmatrix}_{1 \times (K \times T)}$$
(Time-varying)   (3.4)

$$\mathbf{u}' = \begin{bmatrix} u_1 \ u_1 \ \cdots \ u_1 \ \cdots \ u_k \ \cdots \ u_k \ \cdots \ u_k \ \cdots \ u_K \ \cdots \ u_K \end{bmatrix}_{1 \times (K \times T)}$$
(Time-invariant)

$$\mathbf{v}' = \begin{bmatrix} v_{11} \ v_{12} \cdots v_{1T} \cdots v_{k1} \cdots v_{kt} \cdots v_{kT} \cdots v_{K1} \cdots v_{KT} \end{bmatrix}_{1 \times (K \times T)}$$
(Time-varying)   (3.5)

$$\mathbf{v}' = \begin{bmatrix} v_1 \ v_1 \cdots v_1 \ \cdots \ v_k \ \cdots \ v_k \ \cdots \ v_k \ \cdots \ v_K \ \cdots \ v_K \end{bmatrix}_{1 \times (K \times T)}$$
(Time-invariant)

3.3   Data envelopment analysis

3.3.1   Introduction to DEA

Almost all of the work in 'Farrell efficiency' has been restricted to single output situations, although Farrell (1957) does suggest how his exposition can relatively easily be extended to solve a model with multiple outputs. In fact, Farrell's initial work was significantly advanced through the introduction of the first DEA model; the 'CCR' model, named after the authors of the seminal paper in which it was first proposed (Charnes, Cooper and Rhodes, 1978).

DEA can be roughly defined as a non-parametric method of measuring the efficiency of a firm with multiple inputs and/or multiple outputs. This is achieved by constructing a single 'virtual' output to a single 'virtual' input without pre-defining a production function. As has been said, the term DEA and the CCR model were first coined in 1978 (Charnes *et al.*, 1978) but, in the ensuing decades and still so even now, have been followed by a phenomenal expansion of DEA in terms of its theory, methodology and applications. The great influence of the CCR paper is reflected by the fact that by 1999 it had been cited over 700 times (Forsund and Sarafoglou, 2002).

DEA is widely acclaimed as a pivotal technique for measuring efficiency and production possibilities; objectives that are deemed to be one of the common interests of Operational Research and Management Science (Charnes *et al.*, 1994). It is not the intention of this work to provide a comprehensive review of the development of DEA. Not only is there a mismatch between the huge body of DEA literature and the limited space available herein, but also the major focus of this work lies with the application of DEA to the container port industry, and as such, only the key issues relevant to this context are addressed. For a better appreciation of the development of DEA, however, readers are referred to Charnes *et al.* (1994), Seiford (1996), Sarafoglou (1998), and Forsund and Sarafoglou (2002).

3.3.2   DEA approaches using cross-sectional data

DEA is concerned with the efficiency of the individual firm. This firm can be defined as the *unit of assessment* (Thanassoulis, 2001) or the DMU (Charnes *et al.*, 1978) that is responsible for controlling the process of production and making decisions at various levels that may include daily operation, short-term tactics and long-term strategy. DEA is used to measure the efficiency of a firm by comparing it with other homogeneous units that transform the same group of measurable positive inputs into the same types of measurable positive outputs. The basic principle of utilising DEA to measure the efficiencies of firms can be explained conceptually through the use of the example presented in Table 3.2 and Figure 3.3. These show the production of eight container terminals. The productivity of each terminal is the 'throughput/stevedore' in Table 3.2. This is also the slope of the line connecting each point to the origin in Figure 3.3 and corresponds to the number of containers moved per stevedore per unit time. It is clear that T2 is the most efficient compared with the other points. As such, the line from the origin through T2 defines the production frontier. All the other points are inefficient compared with T2 and are 'enveloped' by the production frontier. Their relative efficiencies in the context of DEA, as shown in the bottom line of Table 3.2, are measured by comparing their productivity with that of T2. The term 'Data Envelopment Analysis' stems directly from this graphic description of

Table 3.2   Single input and single output

Terminal	T1	T2	T3	T4	T5	T6	T7	T8
Stevedores	10	20	30	40	50	50	60	80
Throughput	10	40	30	60	80	40	60	100
Productivity (Throughput/Stevedore)	1	2	1	1.5	1.6	0.8	1	1.25
Efficiency (%)	50	100	50	75	80	40	50	62.5

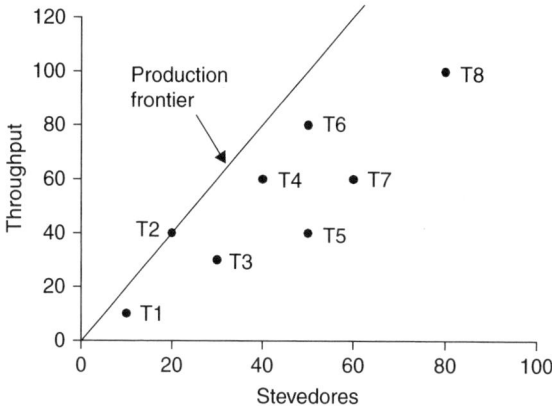

Figure 3.3 Comparison of efficiencies of container terminals (CCR model).
Source: Drawn by the authors.

the frontier 'enveloping' datapoints and of datapoints being 'enveloped' by the frontier.

A more scientific approach to measuring the efficiencies of firms with multiple inputs and outputs is the CCR model (Charnes *et al.*, 1978).

The CCR model for the standard notation employed within this work and previously expanded in Section 3.2 can be expressed by equations (3.6)–(3.9):

$$(\text{FP}_k) \quad \text{Max} \quad U_k = \frac{\displaystyle\sum_{n=1}^{N} a_n y_{nk}}{\displaystyle\sum_{m=1}^{M} b_m x_{mk}} \tag{3.6}$$

Subject to:

$$\frac{\displaystyle\sum_{n=1}^{N} a_n y_{nk}}{\displaystyle\sum_{m=1}^{M} b_m x_{mk}} \leq 1 \quad (k = 1, 2, \ldots, K) \tag{3.7}$$

$$a_n \geq 0 \quad (n = 1, 2, \ldots, N) \tag{3.8}$$

$$b_m \geq 0 \quad (m = 1, 2, \ldots, M) \tag{3.9}$$

Given the data X and Y in (3.2), the CCR model measures the maximum efficiency of each firm by solving the fractional programming (FP) objective function specified in equation (3.6), where the input weights $a_1, a_2, \ldots, a_N$ and output weights $b_1, b_2, \ldots, b_M$ are parameters to be estimated. k in the objective function (equation 3.6) varies from 1 to K. This means that there are K optimisations for all K firms. Constraint (3.7) reveals that the ratio of 'virtual output' $\sum_{n=1}^{N} a_n y_{nk}$ to 'virtual input' $\sum_{m=1}^{M} b_m x_{mk}$ cannot exceed 1 for each firm. This feature conforms to the economic assumption that the nature of production is such that the output(s) cannot be more than the input(s).

The above FP problem expressed in equations (3.6)–(3.9) is equivalent to the following linear programming (LP) formulation given in equations (3.10)–(3.14) (see, e.g. Cooper *et al.*, 2000):

$$(\text{LP}_k) \quad \text{Max} \qquad U_k = \sum_{n=1}^{N} a_n y_{nk} \tag{3.10}$$

Subject to:

$$\sum_{m=1}^{M} b_m x_{mk} = 1 \tag{3.11}$$

$$\sum_{n=1}^{N} a_n y_{nk} \le \sum_{m=1}^{M} b_m x_{mk} \quad (k = 1, 2, \ldots, K) \tag{3.12}$$

$$a_n \ge 0 \quad (n = 1, 2, \ldots, N) \tag{3.13}$$

$$b_m \ge 0 \quad (m = 1, 2, \ldots, M) \tag{3.14}$$

It is important to note that the computation of the above DEA–CCR model by transforming the FP model into an LP formulation has been of great significance for the rapid development and wide application of DEA, and is the main justification for asserting the superiority of DEA over the approach of Farrell (1957). As a long-established mathematical method with various available sophisticated methods of computation and commercially available solution software, LP possesses inherent advantages that make its solution both easier and more feasible.

Apart from the DEA–CCR model, the BCC model (named after Banker, Charnes and Cooper, 1984), as well as the Additive model, are the other two DEA models that are widely studied and applied. The main difference between these two models and the CCR model is that the former allow for a variable returns-to-scale assumption, while the latter is applicable solely to situations where constant returns-to-scale are assumed. Accordingly, the production frontiers in these models are different. Figure 3.4 shows the piecewise linear estimated production frontier for the same example in Table 3.2, but when the BCC model and the

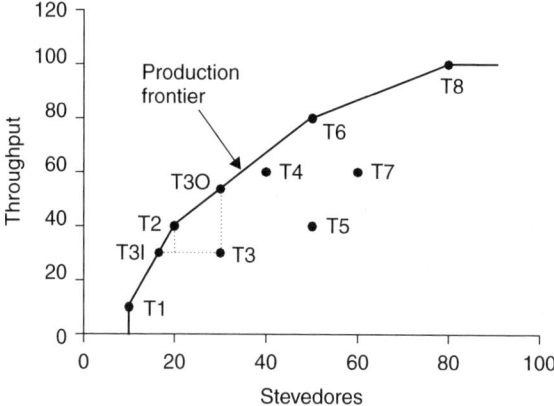

Figure 3.4 Comparison of efficiencies of container terminals (BCC and Additive models).

Source: Drawn by the authors.

Additive model are utilised. The contrast with the production frontier in Figure 3.3, where the CCR model is utilised, can be clearly seen. In Figure 3.4, T1, T2, T6 and T8 on the estimated production frontier are defined as efficient and they cannot dominate each other given the condition of variable returns-to-scale. The other points 'enveloped' by these efficient points are deemed to be inefficient.

The BCC model and the Additive model are identical in terms of their production frontiers. The main difference between them is the projection path to the production frontier from the datapoints that represent the inefficient firms. For instance, the inefficient firm located at T3 can be projected to either T3I or T3O in the BCC model in terms of either the input or output orientation. On the other hand, it is projected to T2 in the Additive model. This different method of projection determines the different estimates of the relative efficiencies for the inefficient firms.

The basic information derived from the above three DEA models, that is, the CCR model, the BCC model and the Additive model, is whether or not a firm can improve its performance relative to the set of firms to which it is being compared. A different set of firms is likely to provide different efficiency results because of the possible movement of the production frontier.

The lack of allowance for statistical noise is widely regarded as the most serious limitation of DEA (Ray, 2002), because this puts a great deal of pressure on users of this technique to collect data on all relevant variables and to measure them accurately. The following two issues, however, have been frequently ignored despite their importance.

The 'virtual' output and input in the context of DEA, as discussed earlier, are defined as the ratio of the sum of weighted outputs to the sum of weighted inputs. This definition is derived from the engineering concept of total factor productivity (Allen *et al.*, 1997) and naturally raises the question as to whether it makes sense to add together different units of outputs or inputs. In other words, as Farrell (1957) himself questions, how reasonable is it to 'add man and acre or pounds and horsepower'?

Without doubt, it is obviously meaningless to simply 'add man and acre'. The explanation of the weights that are attached to inputs and outputs is the only way to understand the principle underpinning DEA. As a relatively recently developed non-parametric method of measuring the technical efficiency of firms, the original objective of DEA was to measure the performance of a firm without either using prices or specifying an explicit technological relationship between inputs and outputs. As such, in the terminology of mathematical programming techniques, it is argued that shadow prices can be used to evaluate the output bundle produced and the input bundle used (Ray, 2002) and the final ratio of weighted output to weighted input can be understood as the ratio of 'shadow revenue' to 'shadow cost'. On the other hand, these weights can also be explained as the 'rates of substitution' or the 'relative values of variables' (Allen *et al.*, 1997).

Another of the important properties of DEA is that there is no requirement for any *a priori* views or information regarding the assessment of the efficiency of firms. The weights for outputs and inputs are obtained by calculating the DEA models, rather than being input artificially as exogenous parameters. By so doing, it is believed that the data are more likely to 'speak for themselves' (Stolp, 1990) and objectively reflect the 'truth' of the situation.

It is interesting to note that this method of selecting weights has not been very frequently challenged. As pointed out by Allen *et al.* (1997):

> The initial development of DEA by Charnes et al. was followed by a rapid expansion of theory and applications without, however, challenging the fundamental basis of DEA insofar as the flexibility in the selection of weights is concerned.

However, discussions continue as to whether the weights that are estimated by DEA might be quite wrong or misleading because they are likely, to some extent, to be different from prior knowledge and accepted views on the relative values of the inputs or outputs. To overcome this drawback associated with DEA, five solutions are proposed:

- to incorporate prior views on the value of individual inputs and outputs;
- to relate the values of certain inputs and/or outputs;
- to incorporate prior views on efficient and inefficient firms;
- to ensure that the assessed efficiency respects the economic notion of input/output substitution;
- to enable discrimination between efficient units.

3.3.3   DEA approaches using panel data

If limited solely to the analysis of cross-sectional data, DEA involves the comparison of one firm with all other firms that produce during the same time period. As such, the role of time is ignored. This can be rather misleading since dynamic settings may give rise to a seemingly excessive use of resources which are intended to produce beneficial results in future periods. The use of panel data is perceived, therefore, as preferable to that of cross-sectional data because not only does this enable a firm to be compared with other counterparts, it also allows the movement in the efficiency of a firm over some time period to be deduced. In so doing, panel data are more likely to reflect the real efficiency of a firm.

When the role of time is considered, let t denote the point in time when the observation is made and T stand for the total number of time periods observed, then the input and output variables of firm k can be rewritten as $(x_{kt}) = (x_{1kt}, x_{2kt}, \ldots, x_{Mkt}) \in R_+^M$ and $(y_{kt}) = (y_{1kt}, y_{2kt}, \ldots, y_{Nkt}) \in R_+^N$, respectively.

Unlike the practice of cross-sectional data analysis, which compares one firm with all other firms in the feasible data set, the analysis of a set of panel data involves choosing only alternative subsets – termed *reference observations subsets* (Tulkens and van den Eeckaut, 1995) – rather than the full data set, in order to evaluate the efficiency of an individual firm. Tulkens and van den Eeckaut (1995) suggest that each observation in a panel can be characterised in efficiency terms *vis-à-vis* three different kinds of frontiers, specified as (i)–(iii) below. Alternatively, the approach originally initiated by Charnes *et al.* (1985) is described as (iv) below:

(i) *Contemporaneous* – involving the construction of a reference observations subset at each point in time, with all the observations made at that time only. A different reference observations subset can be denoted as:

$$\{(x_{kt}, y_{kt}) \mid k = 1, 2, \ldots, K\} \quad \text{for } t = 1, 2, \ldots, T$$

Over the whole observation period, a sequence of T reference observation subsets are constructed, one for each time t.

(ii) *Intertemporal* – involving the construction of a single production set from the observations made throughout the whole observation period. In this case, the reference observations subset is denoted simply as:

$$\{(x_{kt}, y_{kt}) \mid k = 1, 2, \ldots, K; \quad t = 1, 2, \ldots, T\}$$

(iii) *Sequential* – involving the construction of a reference observations subset at each point in time t but, using the observations made from points in time $h = 1$ up until $h = t$. The reference observations subsets at each time $t = 1, 2, \ldots, T$ can be denoted as:

$$\{(x_{kt}, y_{kt}) \mid k = 1, 2, \ldots, K; \quad h = 1, 2, \ldots, t\}$$

Obviously, this method does have the disadvantage of leading to a certain imbalance in the number of observations over which an average efficiency is calculated as t moves towards T. Because of this problem, this approach has been omitted from the analysis contained within this book.

(iv) *Window analysis* – This is a time-dependent version of DEA. The basic idea is to regard each firm as if it were a different firm in each of the reporting dates. Then each firm is not necessarily compared with the whole data set, but instead only with alternative subsets of panel data. Let w be the window width which describes the time duration for the reference observations subsets, then a single window reference observations subset can be expressed as:

$$\{(x_{kt}, y_{kt}) \mid k = 1, 2, \ldots, K; \quad h = t, t + 1, \ldots, t + w; \quad t \leq T - w\}$$

Successive windows, defined for $t = 1, 2, \ldots, T - w$, yield a sequence of reference observations subsets.

Window analysis is based on the assumption that what was feasible in the past remains feasible forever, and the treatment of time in window

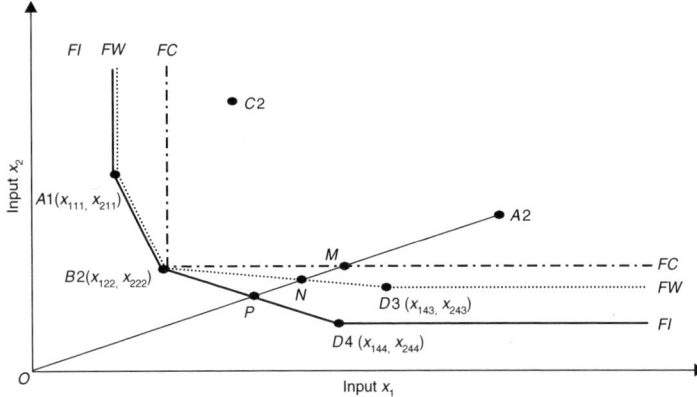

Figure 3.5 Conceptual illustration of contemporaneous, intertemporal and window analyses.

Source: Drawn by the authors.

analysis is more in the nature of an averaging over the periods of time covered by the window. It is difficult to find more than an *ad hoc* justification for the size of the window, as well as for the fact that part of the past is ignored (Tulkens and van den Eeckaut, 1995).

The difference between contemporaneous, intertemporal and the window approaches (where the window width is artificially defined as 3, $w = 3$) is depicted conceptually in Figure 3.5. In this case, an input-oriented model with two inputs that are used to produce the same amount of output(s) is illustrated. Four firms A, B, C and D ($k = 1, 2, 3, 4$) are shown over four time periods ($t = 1, 2, 3, 4$) constituting 16 observations, although not all are mapped in Figure 3.5 (because it is assumed that the unmapped individual observations do not influence the production or cost frontier). In Figure 3.5 then, $D3$ refers to the production of firm D at time $t = 3$. *FI*, *FW* and *FC* in Figure 3.5 stand for the frontiers associated with an intertemporal approach, window analysis and contemporaneous analysis, respectively. $A1$, $B2$ and $D4$ are the most efficient assuming intertemporal analysis and thus together with $(x_{111}, +\infty)$ and $(+\infty, x_{244})$ define the cost frontier for all the panel data (16 observations). However, $D4$ is excluded and replaced by $D3$ in the first window $\{(x_{kh}, y_{kh}) \mid k = 1, 2, \ldots, 4; h = 1, 2, 3\}$. $A1$, $B2$, $D3$ $(x_{111}, +\infty)$ and $(+\infty, x_{243})$ define the cost frontier for the first window. Finally, B2, $(x_{122}, +\infty)$ and $(+\infty, x_{222})$ define the frontier for $A2$ and $C2$ for contemporaneous analysis when $t = 2$. The efficiency of $A2$ yielded by contemporaneous (U_c), window (U_w) and intertemporal (U_i) analyses are *OM/OA2*, *ON/OA2* and *OP/OA2*, respectively. The efficiency of other observations can be derived analogously.

It is clear that $U_i \leq U_w \leq U_c$. in essence, contemporaneous and intertemporal analyses are two extreme situations of window analysis where $w = 1$ and $w = T$. This implies that the efficiency of every observation tends to decline as window width (w) increases. This can be rather simply explained by the fact that a firm in a small sample has less counterparts to be compared against and, therefore, has less chance to be dominated, or alternatively, has a greater chance of being classified as more efficient.

To observe the impacts of window width on efficiency estimates is of great significance. This can be explained by the relationship between total factor productivity (TFP) growth and its two components: technical progress and changes in technical efficiency. Technical progress refers to the shifting of the production frontier over time, while technical efficiency indicates the ability of firms to follow best-practice techniques in order to operate on the frontier at any point in time. Alternatively, technical efficiency involves the analysis of whether the gap between best-practice techniques and actual production methods has been diminishing or widening over time. It is safe to assume that the technology remains the same during the time period defined by the window width w, at least when w is reasonably small (e.g. $w = 1$). In this case, estimates of technical efficiency measure whether the firm is following best practice at that time. On the other hand, when window width is large, technical efficiency estimates are measuring whether a firm is keeping abreast simultaneously of the latest technology and best practice. The differentiation has crucial policy implications in that it helps to identify the source of any inefficiency. For instance, it would be a waste of precious resources if a firm is simply unable to make efficient use of its existing technology and facilities, but instead attributes its inefficiency to technological shortcomings and erroneously embarks on a programme of technological upgrading that unnecessarily absorbs valuable investment funds that have a high opportunity cost.

Despite the numerous advantages of using panel data rather than cross-sectional data, only a few attempts have been made at applying the above non-parametric models using panel data. Among others, these approaches have been used to study the banking industry (Hartman and Storbeck, 1996), police precincts (Sun, 2002), distribution centres (Ross and Droge, 2002) and the US brewing industry (Day *et al.*, 1995). Such methods using panel data have never, however, been applied to the port or container port industry in any geographical context.

3.4 Free disposal hull analysis

As another important non-parametric technique for measuring the efficiency of firms with multiple outputs and inputs and a counterpart of DEA, FDH first appeared in Deprins *et al.* (1984). Because of the conceptual proximity, in this section the FDH approach is presented with reference to the similarities to, and differences from, the DEA approach.

As two deterministic non-parametric methods, DEA and FDH assume no particular functional form for the boundary and ignore measurement error. Instead, the best practice technology is the boundary of a reconstructed production possibility subset based upon directly enveloping a set of observations. These extremal methods use mathematical programming techniques to envelop the data (in a piecewise linear way) as tightly as possible, subject to certain production assumptions that are maintained within the mathematical programming context. FDH assumes strong input and output disposability, with the former referring to the fact that any given level of output(s) remains feasible if any of the input is increased, whereas the latter means that with given input(s) it is always possible to reduce output(s).

DEA adds convexity to the assumptions maintained by FDH. Convex non-parametric frontiers in the context of DEA allow for linear combinations of observed production units. According to this definition, all linear combinations of observations *A* and *C* are feasible in Figure 3.6.

Under such circumstances, the FDH efficient unit *B* is not efficient any more because it is dominated by the new boundary. Figure 3.6 illustrates

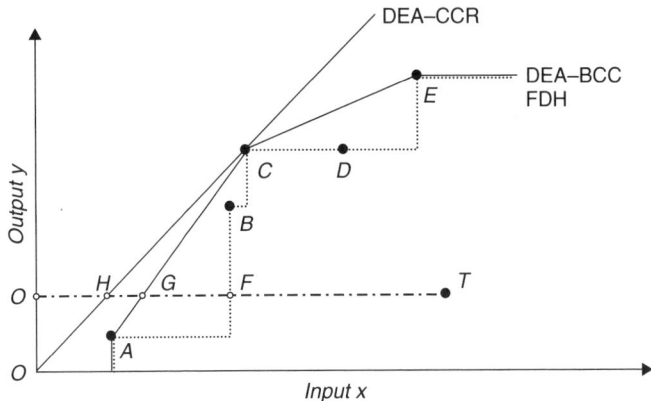

Figure 3.6 Non-parametric deterministic frontiers.
Source: Derived from De Borger *et al.* (2002).

two of the most widely used DEA models, as presented in Section 3.3. The DEA–CCR assumes constant returns to scale so that all observed production combinations can be scaled up or down proportionally. The constant returns-to-scale DEA frontier is derived simply by the ray through the origin passing through point C. The DEA–BCC model on the other hand, allows for variable returns to scale and is graphically represented by the piecewise linear convex frontier.

FDH, DEA–CCR and DEA–BCC models define different production possibility sets and efficiency results. As an example, the input-oriented efficiency of unit *T* in Figure 3.6 is given by *OF/OT*, *OH/OT* and *OG/OT*, as determined by the FDH, DEA–CCR and DEA–BCC models, respectively.

Formally, let inputs be $x_k = (x_{1k}, x_{2k}, \dots, x_{Mk}) \in R_+^M$ to produce outputs $y_k = (y_{1k}, y_{2k}, \dots, y_{Nk}) \in R_+^N$. The row vectors x_k and y_k form the kth rows of the data matrices X and Y, respectively. Let $\lambda = (\lambda_1, \lambda_2, \dots, \lambda_K) \in R_+^K$ be a non-negative vector, which forms the linear combinations of the K firms. Finally, let $e = (1, 1, \dots, 1)$ be a suitably dimensioned vector of unity values.

An output-oriented efficiency measurement problem can be written as a series of K linear programming envelopment problems, with the constraints differentiating between the DEA–CCR, DEA–BCC and FDH models, as shown in equations (3.15)–(3.20):

$$\underset{U, \lambda}{\text{Max}} \qquad U \tag{3.15}$$

Subject to:

$$Uy_k' - Y'\lambda \le 0 \tag{3.16}$$

$$X'\lambda - x_k' \le 0 \tag{3.17}$$

$$\lambda \ge 0 \quad \text{(DEA–CCR)} \tag{3.18}$$

$$e\lambda' = 1 \quad \text{(DEA–BCC)} \tag{3.19}$$

$$\lambda_k \in \{0, 1\} \quad \text{(FDH)} \tag{3.20}$$

The different combination of equations from (3.15) to (3.18), (3.15) to (3.19) and (3.15) to (3.20), respectively, form the DEA–CCR, DEA–BCC and FDH models. It is important to note that input-oriented models can be formulated in a similar way. Interested readers may refer to Seiford and Thrall (1990) Ali and Seiford (1993) and Cooper *et al.* (2000) for more discussion on the above models.

According to Lovell and van den Eeckaut (1993), FDH is gradually becoming more popular, although it is less widely studied and used compared with DEA. Some scholars argue that FDH prevails over DEA in terms of 'data fit' (Tulkens, 1993; van den Eeckaut *et al.*, 1993). However, it is fair to say that both DEA and FDH have their respective strengths and weaknesses (Lovell and van den Eeckaut, 1993). As such, a comparative study of these two approaches may provide greater insight into the intricacies of measuring production efficiency. Efforts in this respect include, *inter alia*, the efficiency of municipalities (van den Eeckaut *et al.*, 1993) and the efficiency of retail banking, courts and urban transit systems (Tulkens, 1993).

In principle, FDH efficiency estimation utilising panel data can be illustrated in an identical manner to that of the DEA panel data approaches. As such, it is not necessary to discuss this in more detail here. One important aspect to point out, however, is that it seems that no applications of FDH models utilising panel data have appeared to be reported in the literature to date. This might be because FDH models, even those using cross-sectional data, are not popularly used, or that the greater complexity of the FDH approach using panel data has deterred applications.

3.5 Econometric approaches using cross-sectional data

As an afterthought, Farrell (1957) proposed computing a parametric convex hull of the observed input–output ratios as an alternative to the non-parametric approach. It is interesting to point out that this afterthought has received even more attention than the non-parametric approach that was actually originally advocated (Greene, 1980). Given cross-sectional data, the parametric model can be classified as deterministic or stochastic according to the assumptions underlying the efficiency and random errors. The former attributes all variation between a firm's production and production frontier as inefficiency, whilst the latter allows for the existence of random errors in the production frontier estimate. These two situations are analysed in Sections 3.5.1 and 3.5.2, respectively.

3.5.1 Deterministic frontier econometric approaches

One of the fundamental assumptions underlying the parametric frontier econometric approach is that there is a parametric function linking output and input variables. Different from DEA or other non-parametric models in which multiple outputs are not uncommon, almost all parametric frontier econometric models only consider one output variable.

Same as the notation in the context of DEA or FDH, let there be $k = 1$, $2, \ldots, K$ firms, using inputs $x_k = (x_{1k}, x_{2k}, \ldots, x_{Mk}) \in R_+^M$ to produce output $y_k \in R_+^1$ for firm k. The technical efficiency of firm k is U_k and must be positive. A parametric production frontier function can be written as in equation (3.21).

$$y_k = f(x_{1k}, x_{2k}, \ldots, x_{Mk}, U_k) \tag{3.21}$$

The first step in solving equation (3.21) is to define the functional form. In principle, the chosen functional form largely depends on *a priori* information about the underlying technology. However, most production in practice is so complicated that the precise engineering blueprints are normally difficult to obtain. Instead, the choice of functional form is usually based on its flexibility and statistical properties (Gallant, 1981; Gong and Sickles, 1992).

Among the few production models available up until this point, the Cobb–Douglas production function has received more attention than others, largely because of its simplicity. Following convention, in this book most models that are discussed are also based on the assumption that any relationship that exists between inputs and output follows that of a homogeneous Cobb–Douglas production frontier. This simplifying strategy is also based on the consideration that since various production frontier functions are associated with different solution approaches, it is difficult to find a universal solution method suitable for all sorts of parametric functional models. This simplification greatly reduces the difficulties associated with the ensuing exposition. It must be noted, however, that as with all mathematical models, the Cobb–Douglas production function cannot be employed on a universal basis, and its suitability to a particular set of circumstances must be strictly examined prior to its application (Beattie and Taylor, 1985).

Given the notation introduced earlier, a homogeneous Cobb–Douglas production frontier based on the assumptions that the production of all firms is efficient can be written as equation (3.22). If inefficiency exists, the production frontier and the technical efficiency of firm k is expressed by equation (3.23):

$$y_k = e^{\beta_0} \cdot \prod_{m=1}^{M} x_{mk}^{\beta_m}, \quad k = 1, 2, \ldots, K \tag{3.22}$$

$$y_k = e^{-\beta_0} \cdot \prod_{m=1}^{M} x_{mk}^{\beta_m} e^{-u_k}, \quad u_k \geq 0, \ k = 1, 2, \ldots, K \tag{3.23}$$

In equation (3.23), $e^{-\beta_0} \cdot \prod_{m=1}^{M} x_{mk}^{\beta_m}$ and e^{-u_k} define the production frontier and the efficiency of firm k, respectively. In the meantime, $u_k \geq 0$, ensures that $0 \leq e^{-u_k} \leq 1$ in order that efficiency values fall between 0 and unity.

For the sake of computational ease in practice, equation (3.23) is normally transformed to its logarithmic format by taking the logarithmic value of both sides of the equation, as shown in equation (3.24):

$$\ln y_k = \beta_0 + \sum_{m=1}^{M} \beta_m \ln x_{mk} - u_k, \quad u_k \geq 0, \ k = 1, 2, \ldots, K \tag{3.24}$$

The two objectives of solving any econometric model such as that in equation (3.24) are to obtain estimates of the parameters describing the structure of production technology and to obtain firm-specific estimates of technical efficiency. According to Kumbhakar and Lovell (2000), there are three alternative approaches to solving equation (3.24): mathematical programming, maximum likelihood estimation (MLE) and regression estimators. The first approach calculates the elements of the parameter vector $\beta = (\beta_0, \beta_1, \ldots, \beta_M)$ by a linear programming model (Aigner and Chu, 1968) which minimises the sum of the absolute values of the residuals, subject to the constraint that each residual be non-positive, or by a quadratic programming model (Kumbhakar and Lovell, 2000), which minimises the sum of squared residuals, subject to the same constraint. The technical efficiency of each firm can be calculated by

$$U_k = \exp\left(-\left(\beta_0 + \sum_{m=1}^{M} \beta_m \ln x_{mk} - \ln y_k\right)\right)$$

The primary advantage of this mathematical programming approach is its computational simplicity. This approach, however, has no statistical properties. In other words, mathematical programming procedures produce 'estimates' without standard errors, t-ratios and so on. This weakness makes it less attractive in the field of econometrics as statistical inferences cannot be drawn from the estimates that result from the analysis.

Equation (3.24) can also be estimated by maximum likelihood estimation (MLE) given specific assumptions on x and u. These assumptions normally include the characteristic that u is independently and identically distributed (i.i.d), $\ln x$ is exogenous (i.e. independent of u), and that the distributions for u can be specified (Afriat, 1972; Richmond, 1974; Schmidt, 1976).

It should be emphasised that the choice of a distribution for u is important because the MLE depends on it in a fundamental way – different assumed distributions lead to different estimates. This, however, is a problem in the absence of any *a priori* arguments for any particular distribution.

As pointed out by Schmidt (1976), one of the regularity conditions for MLE is that the range of the random variable in question is independent of the parameters. However, the specific requirement of $u_k \geq 0$ leads to the conclusion that

$$\ln y_k \propto \left(-\infty, \beta_0 + \sum_{m=1}^{M} \beta_m \ln x_{mk} \right)$$

In other words, the dependent variable (output) depends on the parameters to be estimated. This violates one of the regularity conditions and MLE is not consistent with an asymptotically efficient estimator in this case.

The third approach of regression estimators mainly includes 'corrected ordinary least squares' (COLS) and 'modified ordinary least squares' (MOLS) (Afriat, 1972; Richmond, 1974; Lovell, 1993). COLS suggests two steps to estimate the deterministic production frontier model. In the first step, ordinary least squares (OLS) is used to obtain consistent and unbiased estimates of the slope parameters and a consistent but biased estimate of the intercept parameter. In the second step, the biased OLS intercept β_0 is shifted up ('corrected') to ensure that the estimated frontier bounds the data from above. Thus, the COLS residuals $\hat{u}_k^*$ are non-negative, with at least one being zero, and can be used to provide consistent estimates of the technical efficiency of each producer by means of calculating $U_k = \exp\{ - \hat{u}_k^* \}$.

Two potential problems with the COLS technique include the consistency of the estimated parameters (Greene, 1980) and the structure of the production frontier. As far as the latter is concerned, the production frontier estimated by COLS is parallel to the OLS regression, as only the OLS intercept is corrected. This implies that the structure of the 'best practice' production technology is assumed to be the same as the structure of the 'central tendency' production technology. However, the structure of the best practice production technology ought to be permitted to differ from that of the production technology that is deployed in the middle range of the data observations where firms can be interpreted to be less efficient. This is an undesirably restrictive property of the COLS procedure.

MOLS is an alternative estimation method to solve equation (3.24). Letting μ be the mean of u, equation (3.24) can be written as equation (3.25):

$$\ln y_k = (\beta_0 - \mu) + \sum_{m=1}^{M} \beta_m \ln x_{mk} - (u_k - \mu),$$
$$u_k \geq 0, \quad k = 1, 2, \ldots, K \quad \text{(3.25)}$$

The new error term in equation (3.25) has a zero mean. Indeed, the error term satisfies all of the usual ideal conditions except normality. Therefore, equation (3.25) may now be estimated by OLS to obtain the best linear unbiased estimates of $(\beta_0 - \mu)$ and each of the values of β_m. It is clear that the value of β_0 can be estimated if μ is estimated given the estimates of $(\beta_0 - \mu)$. If a specific distribution is assumed for u, such as an exponential or half-normal distribution, and if the parameters of this distribution can be derived from its higher-order (second, third, etc.) central moments, then these parameters can consistently be estimated from the moments of the OLS residuals. Since μ is a function of these parameters, it too can be estimated consistently, and this estimate can then be used to 'modify' the OLS constant term, which is a consistent estimate of $(\beta_0 - \mu)$.

A problem associated with the MOLS technique is that, even after modifying the constant term, some of the residuals may still have the 'wrong' sign so that these observations end up above the estimated production frontier. This makes the MOLS frontier a somewhat awkward basis for computing the technical efficiency of individual observations. Another difficulty with the MOLS technique is that the correction to the constant term is not independent of the distribution assumed for u (Kumbhakar and Lovell, 2000).

3.5.2 Stochastic frontier approaches

To overcome the problems associated with deterministic frontier approaches which do not allow for the possibility of random shocks, Aigner *et al.* (1977) and Meeusen and van den Broeck (1977) simultaneously, though independently, introduced stochastic production frontier models. These models not only allow for technical inefficiency, but also acknowledge the fact that random shocks that are outside the control of producers can affect output (strikes, weather damage, equipment failure, etc.). The essential idea behind the stochastic frontier model is that the error term is composed of two parts. A symmetric component permits random variation of the frontier across firms and captures the effects of measurement error, other statistical 'noise' and random shocks that fall outside the control of the firm. In addition, a one-sided component captures the effects of inefficiency relative to the stochastic frontier. Because of this characteristic of having two components in the error term, the stochastic production frontier model is often referred to as a 'composed error' model. A stochastic frontier model can be expressed as equation (3.26), which is similar to equation (3.21) except for its statistical noise component V_k:

$$y_k = f(x_{1k}, x_{2k}, \ldots, x_{Mk}, U_k, V_k) \qquad (3.26)$$

Similar to the procedure for solving the deterministic model, the first step in solving a stochastic frontier model is to specify a functional form. The difference between the solutions to the two types of model is that the solutions to stochastic frontier models more frequently rely upon maximum likelihood estimation. The other two solution methods (mathematical programming and regression analysis), although technically feasible, are seldom reported as viable means of solving stochastic frontier models.

As shown in equation (3.27), the log-linear Cobb–Douglas form is again assumed in order to illustrate the solutions to the stochastic frontier model:

$$\ln y_k = \beta_0 + \sum_{m=1}^{M} \beta_m \ln x_{mk} - u_k + v_k,$$
$$u_k \geq 0, \quad k = 1, 2, \ldots, K \qquad (3.27)$$

where v_k denotes the two-sided noise component, and u_k is the nonnegative inefficiency component. The maximum likelihood estimation method is used to estimate the values of both β and u.

All maximum likelihood estimation procedures require assumptions to be made about the variables, the inefficiency component and the statistical errors. The four models utilised herein have been hitherto suggested and vary according to the different assumptions each of them makes about the distributions of u_k and v_k. Since it is not the intention of this book to focus on the intricacies of the computational details of MLE and efficiency estimates, readers seeking greater detail are referred to the works of Aigner *et al.* (1977), Meeusen and van den Broeck (1977), Stevenson (1980), Greene (1990) and Jondrow *et al.* (1982). Herein, and as summarised in Table 3.3, only the results relating to the efficiency estimates are provided.

Table 3.3 Efficiency estimates by maximum likelihood*

Models	Assumptions	Efficiency component
The normal–half normal model	• $v_k \sim i.i.d.\ N(0, \sigma_v^2)$ • $u_k \sim i.i.d.\ N^+(0, \sigma_u^2)$ • v_k and u_k are distributed independently of each other and of the regressors	$$E[u_k|\varepsilon_k] = \frac{\sigma\lambda}{(1+\lambda^2)}\left[\frac{\phi\left(\dfrac{\varepsilon_k\lambda}{\sigma}\right)}{\Phi\left(-\dfrac{\varepsilon_k\lambda}{\sigma}\right)} - \frac{\varepsilon_k\lambda}{\sigma}\right]$$
The normal–exponential model	• $v_k \sim i.i.d.\ N(0, \sigma_v^2)$ • $u_k \sim i.i.d$ exponential • v_k and u_k are distributed independently of each other and of the regressors	$$E[u_k|\varepsilon_k] = (\varepsilon_k - \theta\sigma_v^2) + \sigma_v\phi\left[\frac{\left(\dfrac{\varepsilon_k - \theta\sigma_v^2}{\sigma_v}\right)}{\Phi\left[\dfrac{\varepsilon_k - \theta\sigma_v^2}{\sigma_v}\right]}\right]$$
The normal–truncated normal model	• $v_k \sim i.i.d.\ N(0, \sigma_v^2)$ • $u_k \sim i.i.d.\ N^+(\mu, \sigma_u^2)$ • v_k and u_k are distributed independently of each other and of the regressors	$$E[u_k|\varepsilon_k] = \left(\frac{\varepsilon_k\lambda}{\sigma} + \frac{\mu}{\sigma\lambda}\right)$$
The normal–gamma model	• $v_k \sim i.i.d.\ N(0, \sigma_v^2)$ • $u_k \sim i.i.d.$ gamma • v_k and u_k are distributed independently of each other and of the regressors	$$E[u_k|\varepsilon_k] = \frac{h(p+1, \varepsilon_k)}{h(p, \varepsilon_k)}$$ $$h(p, \varepsilon_k) = E[z^p|z>0, \varepsilon_k]$$ $$z \approx N[-(\varepsilon_k + \sigma_v^2/\sigma_u), \sigma_v^2]$$

* $\sigma = (\sigma_u^2 + \sigma_v^2)^{1/2}$, $\lambda = \sigma_u/\sigma_v$, $\varepsilon_k = v_k - u_k$, and $\Phi(\cdot)$ and $\phi(\cdot)$ are the standard normal cumulative distribution and density functions.

Once point estimates of u_k are obtained, estimates of the technical efficiency of each producer can be obtained from:

$$U_k = \exp\{-E[u_k/\varepsilon_k]\} \tag{3.28}$$

Schmidt and Sickles (1984), Gong and Sickles (1989, 1992) summarise the following three difficulties with maximum likelihood estimation of cross-sectional stochastic production frontier models:

(i)   Both maximum likelihood estimation of the stochastic production frontier model and the subsequent separation of technical inefficiency from statistical noise require strong distributional assumptions on each error component.

(ii) Maximum likelihood estimation also requires an assumption that the technical inefficiency error component be independent of the regressors, although it is easy to imagine that technical inefficiency might be correlated with the input vectors that producers select.

(iii) While the technical efficiency of producers can be estimated using the technique suggested by Jondrow *et al.* (1982), it cannot be estimated consistently, since the variance of the conditional mean or the conditional mode of $(u_k \mid \varepsilon_k)$ for each individual producer does not go to zero as the size of the cross-section increases. Efficiency measurement based on cross-sectional data fails, therefore, to identify unconditional firm-specific efficiency.

3.6   Econometric approaches using panel data

Because a panel contains more information than does a single cross-section, the analysis of panel or longitudinal data (repeated observations on each producer over time) is the subject of one of the most active and innovative bodies of literature in econometrics. Consequently, it is to be expected that access to panel data will either enable some of the strong distributional assumptions used with cross-sectional data to be relaxed or, alternatively, that it will result in estimates of technical efficiency with more desirable statistical properties.

Schmidt and Sickles (1984) point out that each of the three limitations discussed in Section 3.5.2 might be overcome if panel data are available. First, having access to panel data enables us to adapt conventional panel data estimation techniques to the technical efficiency measurement problem, and not all of these techniques rest on strong distributional assumptions. In essence, therefore, repeated observations on a sample of firms can serve as a substitute for strong distributional assumptions. Second, not all panel data estimation techniques require the assumption of independence of the technical inefficiency error component from the regressors. Repeated observations on a sample of producers can also serve as a substitute for the independence assumption. Finally, since adding more observations on each producer generates information not provided by adding more firms to a cross-section, the technical efficiency of each firm in the sample can be estimated consistently as $T \rightarrow +\infty$, with T being the number of observations on each producer. Repeated observations on a sample of producers resolves the inconsistency problem by utilising the technique suggested by Jondrow *et al.* (1982), although this final potential benefit of having access to panel data should be accepted with caveats because many panels are relatively short.

Extant research on the estimation of efficiency by using panel data is normally classified according to whether the assumption made is that the technical efficiency is time-invariant or time-varying. The former refers to a model in which technical efficiency is allowed to vary across firms, but is assumed to be constant through time for each firm. On the other hand, the time-varying assumption refers to a model in which technical efficiency is allowed to vary not only across firms, but also is assumed to be varying through time for each firm.

An explicit justification for the assumption of time-invariant efficiency is made by Gong and Sickles (1992) who claim that firm-specific inefficiency can be regarded as an inherent or structural residual between observed data and the corresponding production (or cost) frontier. In the absence of sudden and dramatic changes in economic environments (e.g. market deregulation perhaps), firm-specific efficiency and its relative ranking are not likely to change drastically over short time periods. On the other hand, assuming the time-invariance of technical efficiency over many time periods may be rather a problematic assumption to make, particularly if the operating environment is competitive. For this reason, time-varying efficiency is generally held to be more likely to reflect true underlying trends in efficiency.

In comparison, stochastic frontier models based upon the assumption of time-invariant efficiency have been relatively more frequently applied (but still only occasionally in absolute terms) than those based on time-varying efficiency. This is largely because of the computational complexity of the latter. In recent years, the rapid development of econometric software such as FRONTIER and LIMDEP (Coelli, 1992; McKenzie and Takaoka, 2003) has greatly contributed to the development of solutions to the estimation problems associated with stochastic frontier models that assume time-varying efficiency.

If stochastic frontier models utilising panel data still assume a log-linear Cobb–Douglas form, equations (3.29) and (3.30), respectively, show the time-invariant and time-varying efficiency models. It is clear that these two models are rather similar in form except for the inefficiency component.

$$\ln y_{kt} = \beta_{0kt} + \sum_{m=1}^{M} \beta_m \ln x_{mkt} - u_k + v_{kt},$$
$$u_k \geq 0, \quad k = 1, 2, \ldots, K \tag{3.29}$$

$$\ln y_{kt} = \beta_{0kt} + \sum_{m=1}^{M} \beta_m \ln x_{mkt} - u_{kt} + v_{kt},$$
$$u_k \geq 0, \quad k = 1, 2, \ldots, K \tag{3.30}$$

Production form using the panel data model differs from that using cross-sectional data in that there is an addition of time subscripts to the output, inputs and statistical noise. In the time-invariant efficiency model (equation (3.29)), the efficiency level remains the same throughout the study period for the same firm. This is very similar to a conventional panel data model with producer effects but without time effects. The only difference is that, since they represent technical inefficiency, the producer effects are required to be non-negative. Solutions to the time-invariant efficiency model can be differentiated between fixed- and random-assumptions of the producer effects on the variables.

The time-varying efficiency econometric model (equation 3.30) is similar to its time-invariant counterpart except for the notation for efficiency. In the time-varying econometric model, efficiency is not necessarily the same over time.

Both equations (3.29) and (3.30) assume that the structure of the production frontier remains the same over time because each of the coefficients β_m are assumed not to vary across firms and with time. These assumptions are mainly for the sake of computational tractability and have been extensively studied in the field of econometrics, even though they might be problematic in reality. Time-varying coefficient models have recently been summarised and reviewed in Kalirajan and Shand (1999).

Existing studies also impose different assumptions on the intercept of the production frontier with y-axis β_{0kt}. These different assumptions are particularly important for obtaining solutions to the models.

Unless otherwise specifically stated, in this book the intercept is assumed to be the same for each firm over time, but different across firms.

Solutions to panel data econometric efficiency models are based on the various assumptions imposed on the models, as summarised in Figure 3.7. Time-invariant models are generally divided into fixed- and random-effects models. Within group estimation and least squares dummy variable estimation are the two most frequently used solution methods for the fixed-effects model, whilst generalised least squares estimation (Hausman, 1978; Hausman and Taylor, 1981; Breusch *et al.*, 1989), instrumental variable estimation and maximum likelihood estimation are particularly used for the latter. Time-varying efficiency models are more complicated and are basically differentiated by the different assumptions imposed on the intercept of the production frontier β_{0kt}. Given the complexity of the solutions to panel data econometric efficiency models and the focus of this work being more on the application of these approaches, solution algorithms for panel data econometric efficiency models are not outlined in detail in this book. Interested readers are referred to Kumbhakar and Lovell (2000) and Greene (2002) for more in-depth coverage.

3.7 Empirical comparisons of alternative approaches to efficiency estimation

Over decades of development, efficiency measurement has advanced rapidly in terms of both methodology and applications. Existing theory suggests that different approaches have different strengths and weaknesses and, therefore, it is difficult to conclude that one approach is necessarily superior to another. Despite this, numerous attempts have been made, through either simulation or empirical testing, to shed light on the strengths and weaknesses associated with the various available approaches.

As a traditional method, simulation models can provide profound insights into the accuracy or appropriateness of alternative approaches. Efforts in this aspect include, *inter alia*, Banker *et al.* (1986), Banker (1996) and Gong and Sickles (1992).

Numerous efforts have also been exerted on empirical comparison of different estimation approaches. Ahmad and Bravo-Ureta (1996) examine the impact of fixed-effects production functions *vis-à-vis* stochastic production frontiers on technical efficiency measures by using an unbalanced panel consisting of 96 Vermont dairy farmers over the period 1971–84. The models examined incorporated both time-varying

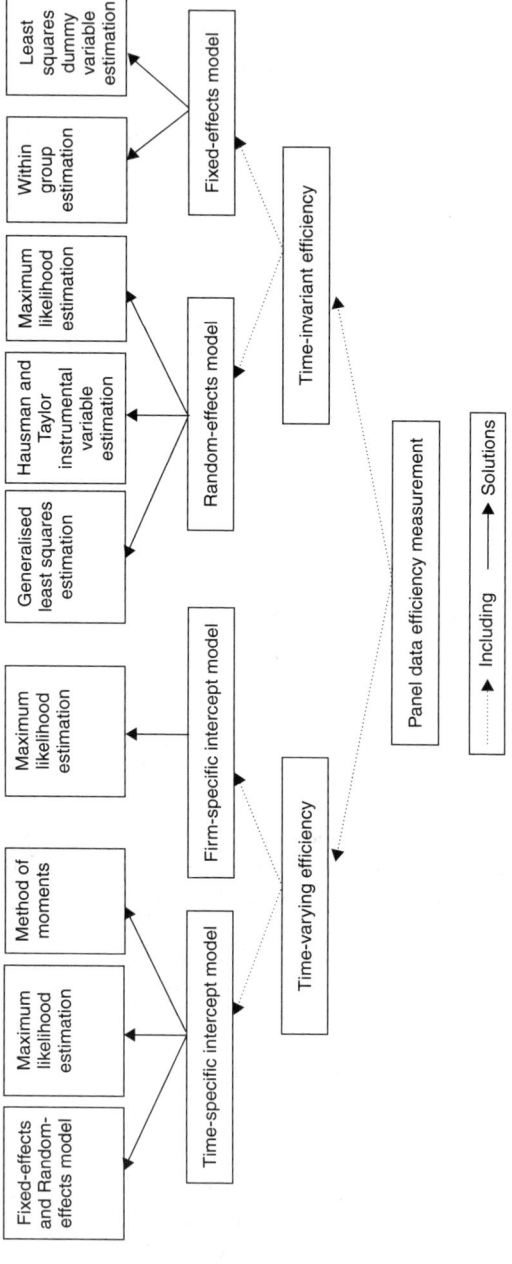

Figure 3.7 Alternative panel data approaches to efficiency measurement.
Source: Drawn by the authors.

and time-invariant technical efficiency. They found that, in general, fixed effects techniques were superior to the stochastic production frontier methodology. Despite the fact that the results of various statistical tests revealed the superiority of some specifications over others, the overall conclusion of the study is that the efficiency analysis was fairly consistent throughout all the models considered.

Banker *et al.* (1986) compared inferences drawn about hospital cost and production from two different estimation models: an econometric model based on the Translog cost function and DEA. Their analysis provided insights into the relative strengths of these two estimation methods by applying both models to the same data. The Translog results suggest that constant returns to scale exists in production, whereas the DEA results suggest that both increasing and decreasing returns to scale may be observed in different stages in the production process, in turn suggesting that the Translog model may be 'averaging' diametrically opposite behaviour.

Bauer and Hancock (1993) examined the efficiency and productivity of the cheque processing offices of the Federal Reserve System using a variety of frontier estimation techniques. They reported that the average level of efficiency varies considerably depending on the technique chosen.

Similar studies can be found in Sherman (1981), Bowlin *et al.* (1985), Sharma *et al.* (1997), Kohersa *et al.* (2000), Ondrich and Ruggiero (2001), Reinhard *et al.* (2000), Odeck (2001), Murillo-Zamorano and Vega-Cervera (2001) and other such. To sum up these studies, it is found that different estimation approaches normally yield different estimates of efficiency. There exists, however, no universal rule or set of conditions for evaluating the applicability of the alternative approaches to efficiency measurement.

3.8   Performance measurement and efficiency analysis in the container port industry

3.8.1   General performance measurement

Traditionally, the performance of ports has been variously evaluated by calculating cargo-handling productivity at berth (Bendall and Stent, 1987; Frankel, 1991; Tabernacle, 1995; Ashar, 1997), by measuring a single factor productivity (De Monie, 1987) or by comparing actual with optimum throughput over a specific time period (Talley, 1998). Suykens (1983) described and discussed the cargo-handling productivity in

European ports over the past several decades. Robinson and Reyes (1998) summarised the development of the efficiency and productivity of ASEAN and Australian ports in the early 1980s.

As discussed in Chapter 2, ports are essentially providers of service activities, in particular, for vessels, cargo and inland transport. As such, it is possible that a port may provide sound service to vessel operators on the one hand and unsatisfactory service to cargo or inland transport operators on the other. Therefore, port performance cannot normally be assessed on the basis of a single value or measure. The multiple indicators of port performance can be found in the example of the Australian port industry (Talley, 1994). The indicators are selected from the perspective of the stevedore, the shipping line and the port authority (or port management). Evaluations are made by comparing indicator values for a given port over time, as well as across ports, for a given time period.

The port performance indicators suggested by the United Nations Conference on Trade and Development (UNCTAD) (1976), as shown in Table 3.4, underlie productivity and effectiveness measures and can be used as a reference point.

Heaver (1995) and Cullinane (2003b) have identified the importance of comparing the performance of one port with its counterparts or its own historical performance. The currently only practical, but rather limited, comparison of the performance of world container ports is the container

Table 3.4 Summary of performance indicators suggested by UNCTAD

Financial indicators	*Operational indicators*
Tonnage worked	Arrival late
Berth occupancy revenue per ton of cargo	Waiting time
Cargo handling revenue per ton of cargo	Service time
Labour expenditure	Turnaround time
Capital equipment expenditure per ton of cargo	Tonnage per ship
Contribution per ton of cargo	Fraction of time berthed ships worked
Total contribution	Number of gangs employed per ship per shift
	Tons per ship-hour in port
	Tons per ship-hour at berth
	Tons per gang-hours
	Fraction of time gangs idle

Source: UNCTAD (1976, pp. 7–8).

port throughput league published in *Containerisation International* and *Port Development International* (Fleming, 1997). Although this container port throughput league can provide some information on the development of the leading container ports in the world, it is unable to reflect the relationship between their inputs and outputs and, hence, can say nothing about their relative efficiency. In consequence, very little useful information can be derived from it.

The approach comparing the productivity of one terminal with that of another, however, has been challenged by some scholars as inappropriate or intractable (e.g. De Neufville and Tsunokawa, 1981). Dowd and Leschine (1990) argued that there is no universally accepted way to compare productivity on a cross-sectional basis. It is more appropriate to compare productivity on a sequential basis, comparing productivity at a single terminal over two or more time periods. Cross-sectional analysis must be conducted with great care, selectively and on a case-by-case basis. This viewpoint is consistent with that of Braeutigam *et al.* (1984) who argued that there are not only various types of ports, but that they are of different sizes and face dealing with great diversity in the traffic mix handled. As such, the use of cross-sectional, time-series or even panel data may fail to show basic differences amongst ports; thus leading to a misjudgement as to each port's performance level. It is crucial, therefore, to estimate econometrically the structure of production in ports at the level of the single port or terminal, using appropriate data such as the panel data for a terminal (Kim and Sachish, 1986). As implied by the discussion contained in Chapter 2, however, this view is particularly questionable given the importance of benchmarking different terminals against one another.

Talley (1994) argued that the criteria specification and the operating objective specification methodologies are two contrasting approaches to the selection of performance indicators in the literature on public transit performance evaluation. The selection criteria include: (1) consistency with goals and objectives; (2) conciseness; (3) data availability; (4) expense and time commitment to collecting indicator data; (5) measurability; (6) minimisation of uncontrollable factors and (7) robustness. The operating objectives of public transit firms have been classified as either effectiveness or efficiency objectives. The former are concerned with the level of service provided by the transit firm to the user, whilst the latter focus on how well the transit firm utilises its available resources. The relationship between these two types of objective means that, according to Talley (1994), a firm must be efficient to be effective.

Talley (1994) went further by attempting to build one single performance indicator – the shadow price variable port throughput per profit
dollar – to evaluate the performance of a port. This overcomes the drawback of multiple indicators, the major one being that examining
whether port performance has improved or deteriorated becomes difficult when changes in some indicators improve performance, while
simultaneous changes in others affect it negatively.

Tongzon (1995) argued that port comparison could only be valid and
meaningful if a port's efficiency is compared with that of a similar port. To
identify 'like for like' ports, two studies were conducted. The statistical
technique of Cluster Analysis was utilised to identify groupings of similar
ports (Tongzon and Ganesalingam, 1994), and Factor Analysis, following
the application of the principal components technique, was applied to
analyse 23 worldwide ports (Tongzon, 1995). In the latter study, the ports
are categorised into three similar groups according to the principal components analysis and on the basis of a weighted average of all five variables analysed: number of gantry cranes, number of TEUs, number of ship
visits, number of TEUs per ship visit and number of container berths.

The study by Tongzon (1995) was strongly challenged by Ashar (1995)
by virtue of the way in which the principle components were defined
and the criteria for categorising low, medium and large principal components. Ashar (1995) argued, in fact, that one may use either the input
or the output side, but not both, to group ports for the purpose of
benchmarking port performance.

3.8.2   Efficiency measurement in the port industry: the DEA approach

As reported by De Borger *et al.* (2002), frontier models (including DEA)
'have found their way to the transport sector, and studies on the
productivity and efficiency of almost all transport modes are now available in the literature'. A comprehensive review of frontier studies on railroads has been conducted by Oum *et al.* (1999). Another more detailed
review of the application of frontier studies to public transit performance
measurement was carried out by De Borger *et al.* (2002).

In recent years, DEA has occasionally been used to analyse port
production. Compared with traditional approaches, DEA has the advantage that consideration can be given to multiple inputs and outputs.
This accords with the characteristics of port production so that there
exists, therefore, the capability of providing an overall evaluation of
port performance. Previous applications of DEA to the port industry are
summarised in Table 3.5. Among the five applications listed, that of

Table 3.5 The application of DEA to ports

References	Objectives of applying DEA	Data description	The DEA model(s) applied	Inputs	Outputs
Roll and Hayuth (1993)	To theoretically rate the efficiency of ports	Hypothetical numerical example of 20 ports	CCR	Manpower Capital Cargo uniformity	Cargo throughput Level of service Users' satisfaction Ship calls
Martinez-Budria *et al.* (1999)	To examine the relative efficiency of ports and efficiency evolution of an individual port	26 Spanish ports using 5 observations for each port during 1993–97	BCC	Labour costs Depreciation charges Other expenditure	Total cargo moved through the docks Revenue obtained from the rent of port facilities
Tongzon (2001)	To specify and empirically test the various factors which influence the performance and efficiency of a port	4 Australian and 12 Other international container ports for year 1996	CCR Additive	Number of cranes Number of container berths Number of tugs Terminal area Delay time Labour	Cargo throughput Ship working rate
Valentine and Gray (2001)	To compare port efficiency with particular type of ownership and organisational structure to determine any relationship with efficiency	31 container ports out of the world's top 100 container ports for the year 1998	CCR	Total length of berths Container berth length	Number of containers Total tons throughput
Barros and Athanassiou (2004)	To estimate the relative efficiency of a sample of Portuguese and Greek seaports	2 Greek ports and 4 Portuguese ports During 1998–2000	CCR BCC	Ships Movements of freight Total cargo handled Containers handled	Labour Capital

Roll and Hayuth (1993) should be treated as a theoretical exploration of applying DEA to the port sector, rather than as a genuine application. This is because no genuine data were collected and analysed. In Martinez-Budria *et al.* (1999), 26 ports were divided into three groups; namely, high complexity ports, medium complexity ports and low complexity ports. After examining the efficiency of these ports using DEA–BCC models, Martinez-Budria *et al.* (1999) concluded that the ports of high complexity are associated with high efficiency, compared with the medium and low efficiency found in other groups of ports. Their findings, however, are particularly dubious given the properties of the DEA–BCC models used in their studies. As further discussed in Chapter 5 (with special reference to Table 5.2 and related discussion), efficiency estimates derived from DEA–BCC models are, to a large extent, influenced by the sampling frame. Some firms are estimated to be efficient simply because they have few similar observations and have less chance to be dominated by other firms. It does not necessarily imply that these firms are inherently better than others.

Using both DEA–CCR and DEA–Additive models, Tongzon (2001) studied the efficiency of 4 Australian and 12 other international container ports for the year 1996. Clearly plagued by a lack of data availability and the small sample size (only 16 observations), Tongzon (2001) identified more efficient ports than inefficient ports. Realising this serious drawback, Tongzon (2001) suggested that further effort should be exerted into collecting more observations to enlarge the sample analysed.

In order to compare port efficiency with a particular type of ownership and organisational structure to determine any relationship with efficiency, Valentine and Gray (2001) applied the DEA–CCR model to 31 container ports out of the world's top 100 container ports for the year 1998. The worrying point that deserves special attention is the definition of input and output variables. In their study, the number of containers and total tons of throughput are chosen as output measures of port production, whilst the total length of berth and average container berth length are defined as input variables. Given the characteristics of port production as discussed in Chapter 2, the appropriateness of this limited definition of input and output variables can be brought into question.

Barros and Athanassiou (2004) apply DEA to the estimation of the relative efficiency of a sample of Portuguese and Greek seaports. The broad purpose of this exercise was to facilitate benchmarking so that areas for improvement to management practices and strategies could be identified and, within the context of European ports policy, improvements implemented within the seaport sectors of these respective countries. The economic benefits of so doing were then evaluated. The results of the analysis point to a particular efficiency shortfall in the port of Thessaloniki and appropriate benchmarks were identified by which this port could improve its performance. Given the high level of average efficiency amongst the whole sample, seeking to achieve the overall objective of scale efficiency was recommended as the paramount objective for the ports under study. In addition, privatisation was advocated as the most appropriate method for achieving economic efficiency.

3.8.3 Efficiency measurement in the port industry: the SFA approach

Table 3.6 summarises the application of SFA to the port industry. Liu (1995) set out to test the hypothesis that public sector ports are inherently less efficient than those in the private sector. A unique set of panel data relating to the outputs and inputs of 28 commercially important UK ports over the period from 1983 to 1990 was collected for analysis. Ownership is hypothesised as a potential factor input to the frontier production function and the effect of its presence on inter-port efficiency differences is investigated by applying the stochastic frontier model to derive the required efficiency estimates. The results revealed that the efficiency difference between ports in the public versus the private sector was negligible and insignificant between trust and municipal ports on the one hand (the public sector) and private ports on the other. Capital intensity is found to have little relationship to efficiency and the impact of size is found to have a small, but significant, impact. Taken together with the scale elasticity that is estimated as part of the analysis, this is not really supportive of the expectation that substantial economies of scale exist within the sector. A peculiarity of this analysis of the UK situation is that port location was discovered to be a significant influence on efficiency, with ports on the west coast of the UK (i.e. on what many shipping commentators have described as the 'wrong' side of the country) being 11 per cent less efficient, on average, than the rest of the sample.

Notteboom *et al.* (2000) applied a Bayesian approach based on Monte-Carlo approximation to the estimation of a stochastic frontier model aimed at assessing the productive efficiency of a sample of 36 European container terminals located in the Hamburg–Le Havre range and in the Western Mediterranean. The data analysed relates to 1994. The robustness and validity of the estimated model was tested by comparing the results with those of four benchmark terminals in Asia (Singapore, Kaohsiung and Hong Kong's MTL and HIT terminals).

Table 3.6 The application of SFA to ports

References	Objectives of applying SFA	Data description	Production function(s) applied	Inputs	Output
Liu (1995)	To test the hypothesis that public sector ports are inherently less efficient than those in the private sector	28 commercially important UK ports over the period 1983 to 1990	Translog	Labour Capital	Turnover
Notteboom *et al.* (2000)	To assess the productive efficiency of some European container ports	A sample of 36 European container terminals located in the Hamburg–Le Havre range and in the Western Mediterranean, and four leading Asian container ports. The data analysed relates to 1994	Bayesian stochastic frontier models	Terminal quay length Terminal area Number of gantry cranes Average numbers of workers per crane The centrality index of the terminal Diversion distance	Container traffic in TEU
Coto-Millan *et al.* (2000)	To estimate the economic efficiency of 27 Spanish ports	Panel data of 27 Spanish ports for the period 1985–89, with a total of 135 observations	Cobb–Douglas Translog	Total costs	Labour Capital Intermediate consumptions
Cullinane and Song (2003)	To assess the success achieved by Korea's port privatisation policies in increasing the productive efficiency of its container terminals	Five container terminals from the UK and South Korea yielding a total of 65 observations	Cobb–Douglas	Labour Capital	Turnover (excluding property sales)
Cullinane *et al.* (2002)	To analyse the administrative and ownership structures of major container ports in Asia	15 container ports or terminals in Asia for the 10-year period from 1989 to 1998 with the exception of Yantian for 6 years from 1993 to 1998. The sample comprises a total of 146 observations.	Cobb–Douglas	Terminal quay length Terminal area The number of pieces of cargo handling equipment	Annual container throughput in TEU

This study found that ten of the terminals, including all four Asian terminals, had efficiency levels of about 80 per cent, with the majority of ports in the Hamburg–Le Havre range falling into the 75–80 per cent efficiency category. The Belgian container terminals were found to be some of the most efficient in the sample, as were the Spanish terminals included in the analysis. Container terminals in Italy were generally found to be the most inefficient. The results also suggested that small terminals tend to be less efficient than large ones, although a more thorough secondary analysis revealed that small terminals located in large ports are, in general, more efficient than small terminals located in smaller ports. Several relationships can be inferred from the results of this study. For instance, situations where intra-port competition is formidable (with no single terminal predominant) seemed to have a positive effect upon average terminal efficiency within the port. In addition, hub ports tended to be more productively efficient than typical feeder ports. However, whether the terminal was publicly or privately owned seems to have no bearing upon the level of efficiency derived. Finally, terminals in the north of Europe are found to be generally more efficient than those in the south of Europe. All of these relationships are investigated, however, on an individual basis and no insight is gained into the potential impact of joint effects. This would appear to constitute, therefore, an interesting avenue for the extension of this research.

Coto-Millan *et al.* (2000) applied a stochastic frontier model to estimate the economic efficiency of 27 Spanish ports. Panel data for the period 1985–89 were collected and analysed using the Cobb–Douglas and Translog versions of the model. The study found that the Translog version of the stochastic frontier model best represents the level of technology within the industry. Quite surprisingly, there appears to be *prima facie* evidence that large ports are found to be rather more economically inefficient than their smaller counterparts, despite the fact that the analysis also reveals the simultaneous existence of significant scale economies within the sector. Indeed, in order of efficiency, the top five positions are occupied by the smallest ports, while the worst five in the efficiency rankings are relatively the largest ports with the greatest degree of autonomy. A more detailed second-stage analysis of the findings, however, reveals that it is not really the size which explains the levels of efficiency that have resulted from the analysis. Rather, it is the degree of autonomy in management that is the critical determining factor, with those ports that are highly autonomous being less efficient than the rest. Technical progress was ascertained to be insignificant over the period covered by the analysis.

Cullinane and Song (2000, 2003) assessed the success achieved by Korea's port privatisation policies in increasing the productive efficiency of its container terminals. The UK container terminal sector provides a useful benchmark for comparison since privatisation and deregulation have formed an integral part of UK port reforms for nearly twenty years and the effect on efficiency, having had time to mature, is much easier to gauge. The stochastic frontier model was justified as the chosen methodology for estimating productive efficiency levels and is applied to cross-sectional data under a variety of distributional assumptions. A panel data model was also estimated. Results were consistent and suggest that: (a) The degree of private sector involvement in sample container terminals is positively related to productive efficiency; and (b) Improved productive efficiency has followed the implementation of privatisation and deregulation policies within the Korean port sector. Even though not categorical, these conclusions are important because the market for container throughput is internationally competitive and if policies which promote competition between Korean container terminals lead to greater productive efficiency, this will inevitably make the sector as a whole more competitive internationally.

Cullinane *et al.* (2002) used the 'port function matrix' due to Baird (1997) to analyse the administrative and ownership structures of major container ports in Asia. The relative efficiency of these ports was then assessed using the cross-sectional and panel data versions of the stochastic frontier model. The estimated efficiency measures were broadly similar for the two versions of the model tested. From the results of the analysis, it was concluded that the size of a port or terminal is closely correlated with its efficiency and that some support exists for the claim that the transformation of ownership from public to private sector improves economic efficiency. It was concluded that while this did provide some justification for the many programmes in Asian ports that aimed to attract private capital into both existing and new facilities, it is also the case that the level of market deregulation is an important intervening variable that may also exert a positive influence on efficiency levels.

3.8.4   Lessons to be learned from applying DEA and SFA to port efficiency measurement

Compared with traditional port performance measurements, the inherent DEA and SFA functions make it possible to capture the overall performance of a container terminal and compare the relative efficiency of different container terminals. The efficiency estimates can provide a benchmark to terminal owners and operators, so that inefficient operators can learn exactly where their shortcomings lie and how, therefore, they might improve their production. In addition, the estimated results can have many policy or managerial implications. For instance, with port privatisation becoming increasingly popular, the estimation of efficiency by DEA or SFA is particularly informative in examining whether privatisation improves efficiency.

By combining the theoretical discussion on DEA and SFA with what has been gleaned from the survey of their applications to the port sector (mainly summarised in Tables 3.5 and 3.6), some broad statements can be made as follows:

• When DEA and SFA are applied, caution is necessary in choosing the firm(s). It is only reasonable to compare different units with similar production functions. In other words, it would be a waste of time to compare a container port with a tanker terminal. Also, most previous studies seem to focus on production at the level of the terminal. This seems to conform to the argument of Alderton (1999) that 'there is little that can be measured on a whole port basis. Most comparable data must concentrate on a terminal basis'.

• Only the technical (in)efficiency of terminals can normally be measured by DEA and SFA, rather than any allocative (in)efficiency. This is because of different port pricing systems, policies and so on. This argument is greatly supported by the fact that most previous studies have tended to focus on technical, rather than allocative efficiency.

• Different researchers have exhibited very little unanimity on what input and output variables should be selected for incorporation into their DEA or SFA models. This is despite the fact that the choices of input and output variables are of great significance for the application of DEA and SFA because 'the identification of the inputs and the outputs in the assessment of DMUs (firms) is as difficult as it is crucial' (Thanassoulis, 2001).

• Almost no identical DEA or SFA models have been selected to analyse different samples. This may imply that the models should be carefully chosen according to the nature of different samples or, say, different sets of firms. As far as DEA applications are concerned, it is argued that, without apparent proof to indicate whether port production follows the economic laws of constant returns to scale or variable returns to scale, both the constant returns-to-scale model (corresponding to the CCR model in the context of DEA) and variable returns-to-scale models (corresponding to BCC and Additive models in the context of DEA) should be considered. The advantage of considering both types of model lies in the fact that the results can provide each firm with information as to what is the likely maximum extent to which its performance can be improved (a projection from the inefficient point to the production frontier in the CCR model), or to what extent it can improve its performance compared with its most similar efficient counterpart (a projection from the inefficient point to the production frontier in BCC or Additive models). As far as SFA applications are concerned, the choice of production function is vital for the appropriateness of the model in reflecting real production. Most existing models that have been applied to the port industry are limited to the homogeneous Cobb–Douglas function or Translog form without *a priori* justification. This can largely be attributed to the limited options of available models and the complexity of production itself. Given the complexity of port production, it seems that these models are not sophisticated enough to reflect real production.

Existing models to explore the efficiency of a container port are based on only a few assumptions defined by the authors. This chapter clearly shows that there are numerous assumptions that might be imposed on the model and that the corresponding efficiency results differ with the different assumptions imposed. As a result, a comparative study between these different models associated with different assumptions constitutes a significant supplementary analysis that is necessary in order to explore the 'true' efficiency of a container port, even though this level of efficiency might never be attainable.

• Panel data are very well suited to analysis using a DEA or SFA model. This is the case even though both these models have been widely utilised in the past to analyse mainly cross-sectional data. In particular, it would be very interesting to undertake such an analysis of panel data to determine whether individual ports become more or less efficient over time and to potentially identify the major reasons behind such a change.

• Given the respective strengths and weaknesses associated with DEA and SFA models, but also of other models, it would be interesting to investigate the robustness of the results achieved by analysing the impact of model specification on the efficiency estimates produced. Such a comprehensive comparative analysis would not only allow an examination of the robustness of each of the alternative modelling methodologies for efficiency estimation, but would also facilitate a profound, and empirically evidenced, insight into the nature of the estimates produced by each model.

3.9 Summary

This chapter has analysed the alternative approaches to measuring efficiency, mainly based on a distinction between non-parametric and parametric approaches. In the case of the former family of approaches, the two leading techniques of DEA and FDH have been compared. Their application using panel data has also been discussed. Parametric approaches, on the other hand, have been found to be more complicated than the former, largely because there exist numerous approaches to their solution that correspond to the various assumptions imposed on the models.

This chapter also discusses alternative non-parametric and parametric model forms with deterministic or stochastic frontiers, and finds that the models with deterministic frontiers generally exhibit the virtue of simplicity. However, they also share a serious deficiency; all variation in output not associated with variation in inputs is attributed to technical inefficiency. The effect of random shocks is not considered even though these sorts of factors, in tandem with others that are under the firm's control (inefficiency), might also contribute (positively or negatively) to variation in output. The advantages of models that revolve around a stochastic frontier lie with the fact that they incorporate stochastic influences that are more realistic in practice. However, to decompose an error into its constituent inefficiency and stochastic elements has to rely upon strong assumptions in relation to the statistical distribution of each of these elements. Without an *a priori* justification of such assumptions, the efficiency estimates that are ultimately derived are not always very convincing.

Following the elaboration of the various approaches to efficiency measurement, this chapter also reviews their application to the port or container port industry. In so doing, the contrast between the potential for their deployment and their actual usage is highlighted. This study shows that existing approaches and techniques for efficiency measurement in the port industry are far from adequate and that, therefore, alternatives should be further and comprehensively explored.

It is well recognised that both parametric and non-parametric approaches have their strengths and weaknesses. However, there remains a lack of empirical proof in relation to their comparative effectiveness in application to the port industry, particularly to the container port industry. The chapter concludes that such a comparative study is not only necessary, but would yield important results and conclusions.

4
Model Specification and Data

4.1 Introduction

This chapter focuses upon the definition of the output and input variables that are required for developing a precise model specification, for the task of data collection and for estimating the efficiency measures for the sample from the port sector. Specifying erroneous or ill-defined variables for collection and analysis will inevitably lead to the wrong conclusions emerging, however elaborate the models employed may be. Sections 4.2–4.4 apply the performance measurement theory suggested by Norman and Stoker (1991) as guidance and utilise the existing literature as a basis for variable justification. DEA model orientation and the specification of the alternative functions for the testing of the parametric models are then discussed in Section 4.5. The detailed procedure to be employed for data collection is reported in Section 4.6 and, finally, Section 4.7 summarises this chapter.

4.2 A procedure for efficiency measurement

Norman and Stoker (1991) suggest several steps for the implementation of efficiency measurement, as summarised in Figure 4.1. The most pertinent of these steps can be encapsulated as follows:

- To identify and define the role and objectives of the firms in the sample population. This involves the identification of where authority lies (and its limits), where responsibility lies (and its limits) and what resources (labour, capital, knowledge, etc.) are at the disposal of the firms. The role must be determined in the context of the role of the whole organisation or service. The important questions in this

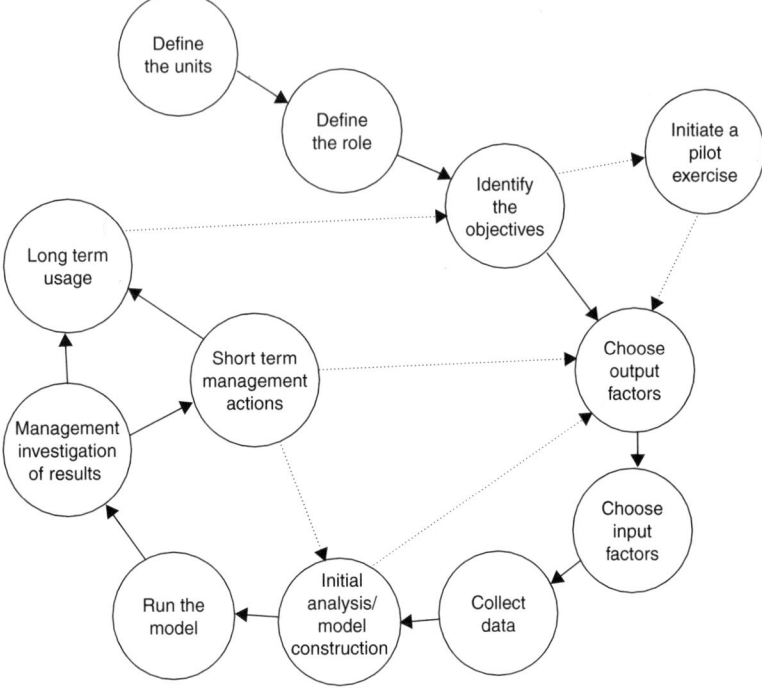

Figure 4.1   A performance measuring system.

Source: Norman and Stoker (1991, p. 168). Copyright John Wiley and Sons Limited. Reproduced with permission.

respect include why was this firm/organisation/unit set up, what does it do and whom does it serve? The definition of the role of the firm or organisation, leads in turn to the identification of a range of objectives that must be achieved.

- The second step involves 'conducting a pilot exercise' and is mainly concerned with the number of firms to be measured. This issue is important because a small number of firms are more likely to generate a high proportion of efficient units, thus robbing the analysis of the opportunity to provide the rich variety of information that is potentially obtainable. Norman and Stoker (1991) suggest that the minimum number of firms that should be considered is 20.
- The third step is to choose the output variables that reflect the provision of support to the achievement of unit objectives. These will be measurable quantities that refer to aspects of achievement. Norman

and Stoker (1991) point out that the golden rule in choosing output variables is to cover the whole gamut of work that the firms undertake. In certain circumstances, however, this might be quite difficult to achieve. Norman and Stoker (1991) go on to recommend that the customers of the firms should first be identified, with prime consideration being given to two questions: (a) who receives the products/services of the firms, and (b) how can these products/services be measured? Deducing the definitions of the output variables to be collected can be based on the answers to these two questions.

It is important here to emphasise that the relationship between the firm objective(s) and choice of input and output variables is also discussed by Lovell (1993) who suggests that only the useful input(s) and output(s) should be considered. If this were not the case, and all possible inputs and outputs were to be considered, then all firms would achieve the same unitary efficiency score since the model would then be overspecified. It is clear that 'usefulness' is closely related to the objective of the firm. One product might be precious for one firm on the one hand and completely useless for another. Stigler (1976) observes that measured inefficiency may be a reflection of a failure to incorporate the right variable and the right constraints and to specify the appropriate economic objectives of the firm. An empirical proof is provided by Kopp *et al.* (1982), who demonstrate the adverse consequences for efficiency measurements of the failure to account for residual discharge constraints in steel production.

- The fourth step is to choose input variables. Norman and Stoker (1991) recommend the normal practice in this respect; that is, to start with a list of factors that is impractically long, then to review the list and to eliminate obvious duplications or irrelevancies.

- The fifth step in data collection is closely related to steps 3 and 4. In many cases, data are not actually available for all input and output variables for which it is desired to collect data. If this happens, Norman and Stoker (1991) recommend three alternative ways of coping with this problem: (a) abridge the list of variables; (b) initiate data gathering and data mining exercises; and (c) a mixture of (a) and (b).

4.3   Port objectives

Port objectives and the functions of ports are closely related and can vary considerably among different ports (Suykens, 1986; Goss, 1990a; Suykens and van de Voorde, 1998). Some traditional port objectives include:

- *Minimisation of user costs*: This includes minimising any or all of the payments by users for the use of the port, the total value (generalised cost) of a ship's time in port (i.e. taking into account the value of a ship's time), or a user's total through-transport costs;
- *Maximisation of benefits*: This includes maximising profits or growth to the benefit of the owners of the port and/or economic benefits to the town, region or country in which the port is located;
- Increasing employment, or 'maximising jobs in the locality'.

The observed performance of a port may be closely related to its objective. For instance, a port is more likely to utilise state-of-the-art and expensive equipment in order to improve its productivity if its objective is simply to maximise cargo throughput. On the other hand, a port may be more willing to use cheaper equipment if its objective is simply to maximise profits.

The objectives of a port form a crucial basis for the definition of variables when undertaking efficiency measurement. Taking the three objectives of port production just mentioned as examples, it is clear that the input and output variables will differ. If the objective of a port is to maximise its profits, for example, then employment or any information on labour should be counted as an input variable. However, if the objective of a port is to increase employment, then employment should be accounted for as an output variable.

As Suykens and van de Voorde (1998) point out, one complication has been that it has been increasingly difficult to precisely define port objectives because of the evolving roles played by different parties involved in container port production. The numerous parties that are now involved in the port industry frequently make it difficult to define clear-cut objectives for ports. Figure 4.2 shows that port management operates under many constraints. In such circumstances, port objectives are likely to be compromised to cater for the requirements of customers, as well as to conform to mandatory regulations.

It is important to note that the roles played by these parties are dynamic and complex. As previously discussed in Chapter 2, even the very large contemporary container ports are sometimes at the mercy of shipping lines, despite the fact that they used to play the dominant role in the market by exercising virtually monopoly control over prices and service quality. This sea change in a port's operating and commercial

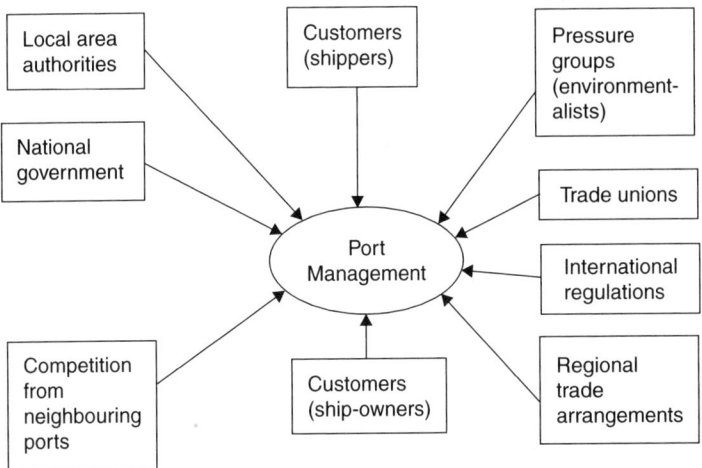

Figure 4.2 Constraining influences on port management.
Source: Alderton (1999, p. 125).

environment has inevitably required individual ports to adjust their objectives so as not to be left behind.

Within the context of this book and in order to facilitate the analysis of container port efficiency, the primary objective for container ports has been assumed to be the minimisation of the use of input(s) and the maximisation of the output(s). This objective is justified by the facts that, *inter alia*, contemporary container ports rely heavily upon sophisticated equipment and information technology (IT) rather than being labour-intensive, as was traditionally the case. The quality of management and the level to which state-of-the-art technology is utilised is hopefully evidence of the achievement of such objectives by individual ports. Similarly, in the light of the fierce and ever-increasing competition that each container port is increasingly facing, achieving this objective is arguably more urgent than any other in order to secure the sustainable development of the port. By way of final justification for this choice of assumed objective, it does also conform to the results of most previous research on the topic.

4.4 Definition of variables

As far as the output from container terminal production is concerned, container throughput is unquestionably the most important and widely accepted indicator of port or terminal output. Almost all previous studies (e.g. Bernard, 1991; Cullinane and Song, 2002) treat it as an output variable, because it closely relates to the need for cargo-related facilities and services and is the primary basis upon which container ports are compared, especially in assessing their relative size, investment magnitude or activity levels.

While Cullinane (2003b) provides a precise distinction between productivity and efficiency measurement, Tongzon (1995) and Brooks (2002) have suggested that a clear distinction is also necessary between port efficiency and effectiveness in order to properly measure container terminal performance. This is a distinction that is more generically addressed by Golany and Tamir (1995). Theoretically, variables that represent the effectiveness of port production should also be incorporated in the model as output variables. Marlow and Paixão (2002) suggest that the following measures of performance would constitute a valid basis for measuring the effectiveness of port services in the supply chain.

1. timeliness in picking up and delivering shipments,
2. reliability of transit time/transport availability,
3. responsiveness of transport suppliers in meeting customers' requirements,
4. adaptability of existing processes to customers' requirements,
5. flexibility of operations,
6. accuracy of information regarding status of shipment,
7. accuracy in processing information,
8. compliance with customers' requirements,
9. value for money,
10. notification of any changes in the multimodal process,
11. level of damage to the shipment,
12. overall transport cost,
13. lead-time to service delivery,
14. level of conflict with other processes,
15. employee interaction with customers.

It is clear that all the measures just mentioned are important contributors to the effectiveness of port production. To incorporate all these variables into a quantitative model of port efficiency would, however, be virtually an almost impossible task, not only because of the data collection problems involved, but also because of the analytical intractability that such complexity would create. An ideal compromise solution, however, is to find a few indicators that provide the closest quantitative proxies to the above performance characteristics and to incorporate these into the analysis.

To this end, the ship working rate (defined as the number of containers moved per working hour per ship) is proposed by Tongzon (2001) as an indicator of the speed with which ships are worked and as a suitable proxy to represent the effectiveness of port service. Since the container handling aspect of port operations is the largest component of total ship turnaround time, the speed of moving cargoes off and onto ships at berth has considerable implications for port users. Moreover, improving efficiency in this area is consistent with port authority intentions of maximising berth utilisation, a factor which will influence both port charges imposed on shipowners and the actual throughput handled. In sharp contrast, Notteboom *et al.* (2000) point out that a container terminal might be highly efficient in terms of the container handling rate per hour, but this does not necessarily imply that, in general, all production factors are used efficiently to generate output. Theoretically, it may well be possible that a seemingly highly productive container terminal in terms of handling rate is only occupied by ships for 10 per cent of its available total operating time.

Ultimately, container throughput has been chosen as the most appropriate and analytically tractable indicator of the effectiveness of a port's production. This is justified on the basis of the theoretical argument that an effective container terminal can attract more clients and that a reasonably high positive correlation is posited, therefore, between increased container throughput and the effectiveness of port production.

There has been even greater debate over the choice of input variables in analyses of this sort. In attempting to derive a port production function, Chang (1978) focuses on general cargo-handling as a measurement of port performance and assumes that port operations follow the conventional Cobb–Douglas case. Chang (1978) argued that the inputs of a port should include the real value of net assets in the port, the number of labourers per year and the average number of employees per month each year, as well as taking into account technological improvement. De Neufville and Tsunokawa (1981) use unloading facilities and space (measured by the number of cranes and the length of quay) as inputs to measure the productivity and returns to scale of container port production. Bendall and Stent (1987) improve the work of Chang (1978) to aid policy makers in assessing the merits of different ship types.

Dowd and Leschine (1990) argue that the productivity of a container terminal depends on the efficient use of labour, land and equipment. Terminal productivity measurement, therefore, is a means of quantifying efficiency in the utilisation of these three resources. Cullinane and Song (2003) suggest that the labour and capital costs of a port or terminal together comprise most of the total cost structure of port or terminal operations. As such, they are sufficient to describe the whole cost account. In Cullinane and Song (2003) *labour inputs* are defined as two different aggregates; one is the total remuneration of directors or executives for their managerial services and the other is the total wages and salaries paid to employees. The *capital input* variable is also divided into two aggregates. One is the net book value of fixed equipment, buildings and land utilised for the purpose of terminal operations and the other is the net book value of mobile and cargo handling equipment including container cranes, yard tractors and fork lifts.

Under the orthodox microeconomic framework, capital and labour costs should necessarily be incorporated in the model. Capital includes the investment made in various port services. UNCTAD (1995) summarises the basic facilities and services a port should provide, as has already been shown in Table 2.1 in Chapter 2.

As with almost all previous studies, the model presented within this book incorporates variables representing all the necessary facilities involved in the container production process (including berths, terminal yardage and equipment) as inputs to the terminal production process and, therefore, as input variables in the efficiency models that are estimated. The first facility to be included as a production input is the container berth. This has been incorporated in a variety of ways into previous applications of the DEA technique to port production research. For instance, Cullinane *et al.* (2002) and Notteboom *et al.* (2000) define total berth length as an input variable. Tongzon (2001), on the other hand, uses the number of berths as an input variable.

Equitable comparability is a paramount criterion for performance measurement (Vancil, 1973). From the perspective of equitable treatment, it may not be appropriate to count the number of berths rather than to count the total length of all berths. Clearly, this is because the number of berths can be changed easily according to market requirements by reconfiguring the quays within a port or terminal and, therefore, is quite an artificial metric. For instance, the same quay of 610 m in length at Yang Ming terminal in Los Angeles was divided into two berths in 1999 and into four berths in 2000. Another drawback in counting simply the number of berths is that this bears no underlying relationship to capacity. Two extreme examples can be found in container terminals in Manila and Gioia Tauro. The length of one berth in Gioia Tauro is 3011 m compared with 318 m for seven berths in Manila. Focusing solely on the number of berths will naturally lead to the conclusion that the single container terminal in Gioia Tauro is more efficient than its counterparts in Manila.

Inequitable treatment that has introduced bias into the estimates of production efficiency in previous studies also exists in the way that terminal equipment has been incorporated into models. The number of gantry cranes in terminals is normally treated as an input variable (e.g. Notteboom *et al.*, 2000). This may be problematic because on the one hand, quayside gantry cranes and yard cranes should be distinguished according to their different functions and usage. On the other hand, the gantry crane is not the only form of equipment that plays a part in container terminal production. In addition, straddle carriers, mobile cranes, front-end handlers, reach stackers, top lifters and fork lifts are also utilised in certain container terminals. One solution is to simply count the aggregate number of all types of equipment present within a container terminal or port. However, problems immediately arise concerning comparability and equitable treatment. For instance, the capacity of just one yard crane is much more than a top lifter. Consequently, a container terminal with more yard cranes will have a higher level of estimated efficiency, even though this high efficiency does not reflect its real input levels.

The solution adopted in this book has been to focus solely on the most important items of equipment for container handling. Yard gantry cranes (including rubber-tyred gantry cranes (RTGs) and rail-mounted gantry cranes (RMGs)), as well as straddle carriers handle the majority of containers in a container yard. As such, it is both reasonable and feasible to treat the absolute number of these separate items of equipment that are operated within a container yard as input variables and to ignore the other items of equipment or facilities that may be deployed within a container yard. A slight exception to this policy is made in the case of some mobile cranes. During the process of data collection, some mobile cranes were found to have quite significant capacities (e.g. over 80 tonnes). In such cases, these mobile cranes have been treated as equivalent to yard gantry cranes because they are capable of handling a similar volume of containers.

Quite apart from terminal facilities (capital), according to the orthodox production theory espoused in mainstream elementary economics, labour should also be incorporated into any model of an industry's production function. Previous research has taken two basic approaches to achieving this. The first and, seemingly, most straightforward is to directly count the number of stevedores and other forms of employee that work in the terminals (see, e.g. Tongzon, 2001; Cullinane and Song, 2003). The disadvantage of this approach, however, is the difficulty of obtaining data and the potential for measurement error. Valentine and Gray (2001) address the inaccuracy of manpower information and explicitly state that, in undertaking their study, information was particularly difficult to obtain from ports that were joint ventures between public and private sector companies. Two extreme examples they provide relate to the port of Melbourne in Australia and the port of Mumbai in India. With a throughput of around 42 million tonnes, the former states that it employs 81 personnel, while the latter, which has a throughput of 30 million tonnes, claims to employ over 24,000 personnel. Taken at face value, this would obviously imply that it takes 300 people in India to do the same job as one person in Australia. Such a stark difference leads to the conclusion that the way the figures are computed by the two different ports must be different.

An alternative approach is to incorporate labour information into the model indirectly. For instance, Notteboom *et al.* (2000) point out that expert analysis has revealed that a fairly stable and close relationship exists between the number of gantry cranes and the number of dock workers in a container terminal (administrative and commercial staff excluded). As such, labour information can be determined purely as a mathematical function of the facilities of a port.

Although the ideal situation would be to incorporate information on port labour directly into the model, this data is both difficult to obtain and often unreliable, either from secondary or even primary sources. Instead, to investigate the relationship between terminal equipment and the number of stevedores, a questionnaire survey was designed and sent to 30 of Asia's leading container terminals. Probably because of the commercial sensitivity of the information that was sought (Robinson, 1999) only 12 useful replies were received. Eight out of the 12 respondents report that the average number of workers per crane is six: one crane operator, one supervisor and four lashing workers. Two container terminals report that they need eight workers for a crane: one foreman, one checker, five stevedores and one crane operator. The final two container terminals report that they needed 15 and 17 workers respectively, for a crane and did not identify the duties of each worker.

Although the sample is admittedly small in terms of absolute responses, the consistency of the responses does tend to support the proposition that the different leading ports use a common labour arrangement. This might be explained by the fact that most important container equipment such as quay gantry cranes or yard gantry cranes are produced by just a few major manufacturers such as Noell in Germany and Zhenhua in China (*Containerisation International Yearbook*, 1998).

In the light of the unavailability and/or unreliability of direct data, information on labour inputs is derived, therefore, from a predetermined relationship to terminal facilities. Apart from the reasons outlined above, another plausible justification is that in this era of modern sophisticated technology and equipment, hackneyed ideas relating to ports being labour-intensive and employing large gangs of basically unskilled dock workers are quite simply completely outdated. Modern technology has greatly reduced the need for the use of labour in port production, especially and most obviously in the developed world and in specialised container ports and terminals. It should be emphasised, however, that in making these assertions and the assumptions that derive from them, there are a number of caveats that should be made explicit. For instance, with the rapid development of manufacturing technology, new products (such as driverless automated guided vehicles) need less labour in absolute numbers. Also, automatic stacking cranes have already been employed at the Delta Dedicated West Terminal of Rotterdam port. Thus, any predetermined relationship between terminal facilities and the absolute number of stevedores employed is not a static one; the derivation of a historic relationship between these characteristics may, therefore, not be reflected in the future; probably as the result of the reaping of even greater efficiency gains. At the same time, it should also be recognised that there is sometimes a dramatic difference in the use of labour in ports of different sizes, in ports where different, usually less sophisticated facilities, are available or even where different types of client are served.

4.5 Model specification

DEA models can be categorised according to whether they are input- or output-oriented. Both orientations have their usefulness within the context of the container port industry. The former is closely related to operational and managerial issues, whilst the latter is more related to port planning and strategies. As far as input-oriented models are concerned, as discussed earlier in Chapter 2, the port industry is normally associated with long-lived assets and a long-term planning horizon and once the port is built, its output is roughly fixed within a certain range for some considerable time. At the same time, a port is normally able to approximately predict its container throughput for the ensuing year at least. This is because a container port has a fairly stable customer base of shipping lines. Over the fairly short-term, container terminals should even be able to predict impending dramatic changes, such as Maersk–Sealand's decision to move its regional hub from Singapore to the Port of Tanjung Pelepas in Malaysia in 2000. A container terminal can also attempt to predict its future throughput by studying historic data or regional economic developments. In these circumstances, how to efficiently use its factor inputs is the key to saving costs or maximising outputs in port production.

As previously mentioned, Marlow and Paixão (2001) argue that leanness and agility are the two key issues for the survival of a port. Leanness requires a port to eliminate all waste including time, while agility attaches great importance to the volatile marketplace and requires the port to be proactive in addressing a changing market. Leanness is a prerequisite for agility. The distinction between leanness and agility provides a useful guide for model choice in terms of whether it should be input- or output-oriented. It is clear that leanness is more closely related to operational matters and as a management strategy is, therefore, easier to implement than agility. All this suggests that an input-oriented model is appropriate to the analysis of container production in a stable market.

On the other hand, with the rapid expansion of global business and international trade, many container ports must frequently review their capacity in order to ensure that they can provide satisfactory services to port users and maintain their competitive edge. Sometimes, the need to build a new terminal or increase capacity is inevitable. However, before a port implements such a plan, it is of great importance for the port to know whether it is fully and efficiently using its existing facilities and that output has been maximised given the input. From this point of view, the output-oriented model will probably provide a better benchmark for the container industry.

Ultimately, it has been decided that an approach based upon the use of output-oriented models should form the basis for the analysis undertaken within this book. There are two fundamental reasons that justify this selection. First, since the main interest of this book lies with informing policy-decisions at national, regional or local levels, an output-oriented model is more appropriate. Second, and from a purely pragmatic point of view, all available alternative models cater for the case where there is a single output. Hence, the choice of an output-oriented approach greatly facilitates the direct comparison of all alternative models on a one-to-one basis. A further attraction lies with the greater analytical tractability and easier data collection that is inherent in adopting just a single output variable as the basis upon which the analysis is undertaken.

The statistical functions of alternative parametric approaches for efficiency estimation are specified by assuming the appropriateness of the log-linear Cobb–Douglas case. For instance, the logarithmic stochastic frontier model for the cases of cross-sectional and panel data are defined by equations (4.1) to (4.3):

$$\ln Y_k = \beta_0 + \beta_1 \ln \text{TL}_k + \beta_2 \ln \text{TA}_k + \beta_3 \ln \text{QG}_k$$
$$+ \beta_4 \ln \text{YG}_k + \beta_5 \ln \text{SC}_k + v_k - u_k,$$
$$k = 1, 2, \ldots, 62 \tag{4.1}$$

$$\ln Y_{kt} = \beta_{0kt} + \beta_1 \ln \text{TL}_{kt} + \beta_2 \ln \text{TA}_{kt} + \beta_3 \ln \text{QG}_{kt}$$
$$+ \beta_4 \ln \text{YG}_{kt} + \beta_5 \ln \text{SC}_{kt} + v_{kt} - u_k,$$
$$k = 1, 2, \ldots, 30; \quad t = 1, 2, \ldots, 8 \tag{4.2}$$

$$\ln Y_{kt} = \beta_{0kt} + \beta_1 \ln \text{TL}_{kt} + \beta_2 \ln \text{TA}_{kt} + \beta_3 \ln \text{QG}_{kt}$$
$$+ \beta_4 \ln \text{YG}_{kt} + \beta_5 \ln \text{SC}_{kt} + v_{kt} - u_{kt},$$
$$k = 1, 2, \ldots, 30; \quad t = 1, 2, \ldots, 8 \tag{4.3}$$

In equation (4.1), Y_k represents the output of the kth container terminal (or port). TL_k, TA_k, QG_k, YG_k, and SC_k denote, respectively, the terminal quay length, terminal yard area, number of quayside gantry cranes, number of yard gantry cranes and number of straddle carriers, associated with the kth terminal. β_0 to β_5 are coefficients associated with the independent input variables in the model and constitute the objects to be estimated. The disturbance term v_k represents the symmetric (statistical noise) component and u_k (≥ 0) is the one-sided (inefficiency) component. The goodness of fit of the estimated regression equation evaluated by R^2 for the least squares method looks reasonably high at 0.8027. This implies that the five inputs to the model do satisfactorily explain the model output. In addition, an F-statistic of 45.56 reveals that the relationship between exogenous and endogenous variables is significant at the 1 per cent level.

Equations (4.2) and (4.3), respectively, correspond to the time-invariant and time-varying efficiency logarithmic stochastic frontier models. It is clear that cross-sectional and panel data modes are similar, except for the absence of the subscript t associated with the variables (such as X and Y) and with the efficiency and random components (u_{kt} and v_{kt}) in the former.

4.6 Data collection

In this book, sets of both cross-sectional and panel data are analysed to allow the estimation of container port or terminal efficiency under a number of differing assumptions and model specifications.

Table 4.1   Summary statistics for sample 1

	Output Container throughput (TEU)	Inputs				
		Terminal length (m)	Terminal area (ha)	Quayside gantry (number)	Yard gantry (number)	Straddle carrier (number)
Mean	2,161,216	3,007	131	21	31	25
Standard Error	324,902	390	21	3	7	5
Median	1,537,178	1,650	80	13	14	2
S.D.*	2,558,284	3,073	167	20	54	39
Kurtosis	14	4	13	4	20	5
Skewness	3	2	3	2	4	2
Range	15,787,793	15,413	995	97	337	171
Minimum	157,000	305	6	2	0	0
Maximum	15,944,793	15,718	1,000	99	337	171
Sum	133,995,403	186,457	8,109	1,272	1,916	1,542
Count	62	62	62	62	62	62

* Standard deviation.

Sample 1: This sample underpins the cross-sectional data analysis and comprises the world's leading container ports ranked in the top 30 in 2001. Out of these 30, the Port of Tanjung Pelepas in Malaysia and San Juan are excluded; the former because it did not officially open until 2000, and the latter because the required data are simply not available. Apart from these world leading container ports, five other container ports in the Chinese mainland are also included although they are not ranked in the top 30 in 2001. Thus, the sample for analysis comprises a total of 62 observations that relate either to individual container ports themselves or to a number of individual container terminals within some of the container ports that are present within the sample. Important descriptive statistics relating to sample 1 are summarised in Table 4.1.

Sample 2: This sample serves as the basis for the panel data analysis and also comprises the world's leading container ports ranked in the top 30 in 2001. Although ranked in the top 30, five container ports (Shenzhen, Gioia Tauro, Tanjung Pelepas, Algeciras and San Juan) are finally excluded from the sample because either they have a shorter history than the study period or lack reliable data for analysis. Apart from these world leading container ports, five other container ports in the Chinese mainland are also included, even though they are not ranked in the top 30 in 2001. Eight years of annual data from 1992 to 1999 are collected for each port. Thus, the sample for analysis comprises a total of 240 observations. The summary descriptive statistics for this sample are reported in Table 4.2.

Table 4.2   Summary statistics for sample 2

	Output Container throughput (TEU)	Inputs				
		Terminal length (m)	Terminal area (ha)	Quayside gantry (number)	Yard gantry (number)	Straddle carrier (number)
Mean	2,455,824	4,609	194	29	39	31
Standard Error	159,069	260	13	2	4	3
Median	1,898,135	3,251	143	23	24	17
S.D.*	2,464,287	4,031	206	25	61	40
Kurtosis	10	1	6	2	14	1
Skewness	3	1	2	1	4	1
Range	15,911,088	16,775	996	110	337	171
Minimum	33,705	142	4	1	0	0
Maximum	15,944,793	16,917	1,000	111	337	171
Sum	589,397,859	1,106,187	46,518	6,950	9,245	7,423
Count	240	240	240	240	240	240

* Standard deviation.

The required secondary data are mainly taken from various issues of both the *Containerisation International Yearbook* and *Lloyd's Ports of the World*. Data on China's container port production are mainly taken from *China's Shipping Development Annual Report* (1998–2000). As the publishers of these sources contact the ports under study every year, and the data are compiled based on their surveys, the data analysed within this study is regarded as the most reliable and comprehensive available.

It is important to note that, during the process of data collection, special emphasis has been placed on the container port industry in China. This is partly because of its rapid development and yet the simultaneously incongruous scarcity of studies on the subject. It is also due, however, to the fact that the fundamental and dramatic changes in China's economic structure that have been instigated in recent years provide a unique context for exploring the nature and veracity of some of the body of economic theory that underpins the workings and structure of the container port industry in general.

Based on the argument that container terminals are more suitable for one-to-one comparison than whole container ports (Wang *et al.*, 2002), this study initially intended to investigate individual container terminals. However, data sources often reported the required data, especially container throughput, at the aggregate level of the whole port, rather than on the basis of the individual terminals that may comprise each of those ports within the sample. In such cases, the input and output of a port were obliged to be defined as the aggregation of the input and output of individual terminals within the port.

4.7   Summary

The scientific definition of input and output variables is the only way to reflect the real process and objectives of production and, therefore, constitutes a keystone of any analysis of efficiency. Mainly because of the numerous inputs involved in container port production to yield the outputs produced, container port or terminal production is well known for its complexity. This chapter has sought to provide a solid definition of, and justification for, the chosen input and output variables in estimating container port or terminal efficiency, giving due consideration to the many and varied objectives that may apply within a port. Representing both the physical cargoes handled in the terminals and the overall (aggregate) effectiveness of terminal production, the container throughput at terminals (and in some cases, in aggregate, at the ports) is chosen as the output variable. The overall length of a terminal's berths, as well as the number of quayside gantry cranes, yard gantry cranes and straddle carriers, are chosen as the input variables because they are the most important facilities for handling containers within a terminal.

Along with the definition of input and output variables, this chapter has also provided the specifications of the models that are to be used for deriving the estimates of port or terminal efficiency. The output-oriented model and log-linear Cobb–Douglas functions are finally chosen as the basis for DEA and parametric analyses, respectively. For both generic forms of model, both cross-sectional and panel data are collected for the estimation of efficiency measures of container port production.

5
Empirical Results and Analysis

5.1 Introduction

This chapter reports the empirical results derived from applying the alternative approaches to container port efficiency estimation discussed in Chapter 3. The efficiency results yielded by various approaches are compared using the Pearson correlation coefficient and Spearman's rank order correlation coefficient in order to obtain some insight into the degree of variation between the results yielded by the various approaches to efficiency estimation.

This chapter is organised according to whether cross-sectional or panel data are used as the basis for the ensuing analysis, each of which sets of tests and results are reported in Sections 5.2 and 5.3, respectively. Section 5.4 summarises and concludes the chapter.

5.2 Cross-sectional data analysis

In this section, sample 1 (i.e. a sample of cross-sectional data) is used to derive the efficiency of container port production. A number of alternative analytical approaches are employed so that the effect of model selection on the results of the efficiency estimation can be examined. Among the various approaches that could be used to analyse the efficiency of container ports that were extensively addressed in Chapter 3, DEA and FDH are the major techniques to be applied from the non-parametric stream, and SFA and COLS are the main ones from the parametric family of approaches.

5.2.1 Non-parametric frontier estimation: DEA versus FDH

The software DEA-Solver-PRO 3.0 (Cooper *et al.*, 2000) is employed to solve both the DEA and FDH models. Without precise information on the nature of the returns to scale within the port production function, two types of DEA models, namely the CCR and BCC models, are applied to analysing the efficiency of the sample of container terminals or ports.

Table 5.1 and Appendix 1 show that average efficiency estimates calculated by DEA–CCR, DEA–BCC and FDH demonstrate an upward trend, with average values of 0.5700, 0.7272 and 0.8880, respectively. To place these figures in their proper context, an index value of 1.0000 equates to perfect (or maximum) efficiency. Eight out of 62 terminals are identified as efficient when the DEA–CCR model is applied, compared with 24 and 40 efficient terminals, respectively, when the DEA–BCC and FDH models are applied. This result is not surprising. As discussed in Chapter 3 and especially in the conceptual illustration shown in Figure 3.6, a DEA model with constant returns to scale provides information on the joint impact of pure technical and scale efficiency taken together. Conversely, a DEA model with variable returns to scale isolates and identifies technical efficiency alone. With $F = 25.22$, an ANOVA of the efficiency estimates for the DEA–CCR, DEA–BCC and FDH analyses indicates that the efficiency measures calculated using these three different approaches are significantly different at the 1 per cent level. The Spearman's rank order correlation coefficients between the efficiency rankings derived from DEA–CCR and DEA–BCC, DEA–CCR and FDH, and DEA–BCC and FDH are 0.8122, 0.6462 and 0.6349, respectively. The positive and high value of Spearman's rank order correlation coefficient indicates that the ranking of each firm in the sample, in accordance with the efficiency estimate

Table 5.1   Summary of the terminal efficiency estimates of the CCR, BCC and FDH models*

	DEA–CCR–O	DEA–BCC–O	FDH
Mean	0.5700	0.7272	0.8880
Standard error	0.0343	0.0343	0.0256
Median	0.5059	0.7570	1.0000
Standard deviation	0.2702	0.2699	0.2017
Sample variance	0.0730	0.0728	0.0407
Kurtosis	−1.1215	−1.0563	2.9584
Skewness	0.3610	−0.4850	−1.9566
Range	0.8460	0.8326	0.7419
Minimum	0.1540	0.1674	0.2581
Maximum	1.0000	1.0000	1.0000
Number of efficient terminals	8	24	40
Count	62	62	62

* 1 = 'Efficient'.

that has been derived by applying the three approaches, is quite similar. A combination of ANOVA and Spearman's rank order correlation coefficient leads to the conclusion that the efficiency estimates yielded by the three approaches follow the same pattern across firms.

Table 5.2 summarises the results on how many efficient terminals result from applying the DEA models and the FDH model. It shows the number of efficient terminals within each class range of production scale as measured by container throughput, and implicitly provides some insight into the rationale behind the use of the two different DEA models and the FDH model. As far as FDH is concerned, 'efficiency by dominating' and 'efficiency by default' are the two sources of efficiency *per se* (van den Eeckaut *et al.*, 1993). The former can be illustrated by the first scale category (throughput of 0–999,999 TEUs) in Table 5.2. Thirteen out of 23 terminals are efficient in the context of the FDH model because they are not 'dominated' by any other terminals. However, within the same scale category, some of them dominate other inefficient terminals. For instance, Terminal 3 of Hong Kong's container port dominates the South Harbour (Manila), Shekou (Shenzhen), Klang Port (Port Klang) and Manila International (Manila). The concept of 'efficiency by default', on the other hand, is demonstrated by observing scale category no. 4 (in the throughput range of 3,000,000–3,999,999 TEUs). In this category, all of the three terminals (Antwerp, Hamburg and Los Angeles) are efficient under the FDH model. A more detailed interpretation of the results reveals that these three ports do not dominate any other ports. They are 'efficient by default', simply because they are not dominated by any other ports.

The relationship between the sample size within different scale categories and the corresponding proportions of inefficient firms is graphically illustrated in Figure 5.1, which is directly derived from Table 5.2. Clearly, Figure 5.1 reveals that, compared with the smaller number of observations in the large throughput (scale) ranges, a higher number of observations in the small throughput size ranges tends to yield higher proportions of FDH and DEA–BCC inefficient terminals. This conclusion, however, does not concur with the results derived from applying the DEA–CCR model. For the DEA–BCC and FDH models, a firm in a small sample (usually associated with higher throughput levels) has fewer counterparts to be compared against (i.e. fewer benchmark container ports or terminals) and, therefore, has less chance of being dominated. Given the assumption of constant returns to scale that is an implicit assumption underpinning the DEA–CCR model, all of the sample firms that fall into the different categories of production scale can be compared directly, with no real

Table 5.2 Summary results on numbers of efficient terminals with DEA models and the FDH model

Category of container throughput (TEU) (1)	Terminals in this category (2)	CCR-I		BCC-I		FDH	
		Efficient terminals (3)	% [5(3)/(2)] (4)	Efficient terminals (5)	% [5(5)/(2)] (6)	Efficient terminals (7)	% [5(7)/(2)] (8)
0–999,999	23	3	13.0	9	39.1	13	56.5
1,000,000–1,999,999	15	4	26.7	5	33.3	10	66.7
2,000,000–2,999,999	13	1	7.7	2	15.4	8	61.5
3,000,000–3,999,999	3	0	0.0	2	66.7	3	100.0
4,000,000–4,999,999	3	0	0.0	2	66.7	2	66.7
5,000,000+	5	0	0.0	4	80.0	4	80.0
Total	62	8	12.9	24	38.7	40	64.5

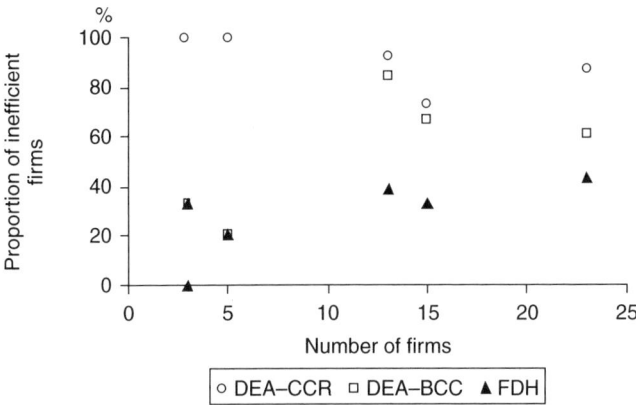

Figure 5.1 Sample size and proportion of inefficient firms. The proportion of inefficient firms is simply calculated by the difference between unity and the proportion of efficient firms.

discernible pattern emerging. This can be explained by the nature of the production possibility set defined by the DEA–CCR model (Cooper *et al.*, 2000). In fact, although undermined by the small sample sizes in each of the different scale categories, it would seem to be reasonable to conclude, on a *prima facie* basis at least, that ports or terminals that are smaller in scale than their counterparts ostensibly appear to be more efficient than their larger competitors.

   In practice, an FDH-efficient firm is not necessarily better than its counterparts with lower efficiency, although an FDH-efficient firm may lose the incentive to improve its production efficiency because it is already efficient in terms of FDH. This is an important drawback of applying the FDH methodology in a management context. To some extent, DEA can overcome this drawback in that it constructs a hypothetical convex hull that envelops the observed firms. In so doing, some FDH-efficient firms may actually be DEA-inefficient. For instance, four FDH-efficient and ten FDH-efficient terminals in the first scale category in Table 5.2 (with throughput of 0–999,999 TEUs) become inefficient when applying, respectively, the DEA–BCC and the DEA–CCR models. The DEA methodology (using either model form) provides a greater opportunity for firms to be benchmarked. For instance, the TCT Tollerort terminal of Hamburg port is FDH efficient but DEA–BCC inefficient. A further investigation reveals that the terminal is benchmarked to be inefficient against a combination of Klang Container Port (Port Klang), Gioia Tauro, Keelung and Nanjing. However, without strong *a priori* support, the constructed convex hull might be too artificial to be convincing and feasible (van den Eeckaut *et al.*, 1993). In this study, the best practice based on the production information of four container ports, Klang Container Port (Port Klang), Gioia Tauro, Keelung and Nanjing, is particularly dubious in terms of its feasibility. For these reasons, a combination of the DEA and FDH analyses can be of great significance in making managerial decisions for ports and terminals and in influencing the strategic decisions of port authorities. On the one hand, the results from applying the FDH model identify the most obvious efficient counterpart(s) from whom inefficient firms can learn in terms of realistically comparable industry 'best practice'. This result is convincing because these efficient counterparts are real. However, the FDH model is more likely to identify as efficient, self-satisfied and undynamic firms that are not really performing that well. In this respect, DEA has the greater potential to provide efficiency aspirations for firms to work towards, although the individual goals that may be set are obviously subject to further study in terms of the feasibility of their being achieved in practice.

Despite the respective strengths and weaknesses associated with the DEA and FDH models, the latter identifies more efficient firms than inefficient firms (40 versus 22), compared with more inefficient firms identified by the DEA. This reflects the fact that the sample data are not sufficiently voluminous to develop a meaningful facet or frontier in the FDH modelling methodology. At the same time, it also suggests that it appears to be more appropriate to use efficiency estimates produced from the application of DEA approaches. For this reason, the FDH model is excluded from analysing panel data in this chapter and from any further discussions of the economic theory underpinning the container port industry that is to be elaborated upon in Chapter 6.

5.2.2 Parametric frontier estimation: COLS versus SFA

In the context of estimating a parametric production frontier, COLS and SFA respectively represent deterministic and stochastic parametric frontier approaches. In order to ensure that MLE is a suitable method of estimation for the solution of the stochastic frontier model, the first step in the estimation procedure is to check the sign of the third moment and the skewness of the OLS residuals associated with the sample data (Waldman, 1982). The skewness for the port production frontier model in this empirical application is -0.5234; the negative sign implying that the residuals of the sample data possess the correct pattern for the implementation of the MLE procedure. This is reflected in the histogram of the residuals shown in Figure 5.2, which is clearly negatively skewed.

In order to obtain the structure of the production frontier in the parametric frontier models, the coefficients of the different variables have been estimated via a range of alternative approaches and the estimation results reported in Table 5.3. It can and should be observed that the MLEs differ only marginally from the OLS estimates and from each other. This is to be expected, since both methods are consistent.

The likelihood ratio (LR) statistic is applied to test whether or not model coefficients are significantly different from zero. Since the maximum likelihood method is a large-sample estimation procedure, an asymptotic test statistic is required to be used. The LR test is one of the general large-sample tests based on MLE and has a χ^2 distribution with degrees of freedom equal to the number of restrictions imposed. According to Engle (1984), under general conditions, the LR test statistic can be expressed by equation (5.1).

$$\text{LR} = -2 \ln \left(\frac{L_R}{L_U} \right) \tag{5.1}$$

where L_R denotes the 'restricted' likelihood function and L_U the 'unrestricted' likelihood function.

The LR test statistics of 102.511, 103.456, 103.657 and 2316.01, calculated in accordance with equation (5.1), correspond to the assumptions

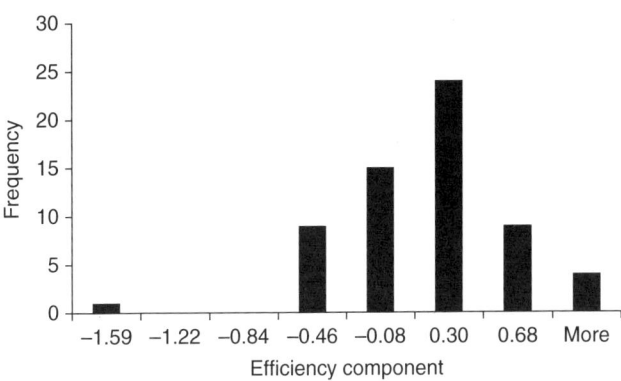

Figure 5.2 Skewness of the OLS residuals for cross-sectional data.

Table 5.3   Frontier production functions of container terminals*

Variables Parameters	COLS**	MLE			
		Half-normal	Truncated-normal	Exponential	Gamma
Constant	11.986	12.628	12.544	12.541	12.512
	(12.747)	(11.265)	(10.067)	(12.495)	(12.330)
ln TL	−0.139	−0.195	−0.212	−0.211	−0.211
	(−0.701)	(−0.820)	(−0.926)	(−0.975)	(−0.980)
ln TA	0.144	0.176	0.184	0.184	0.184
	(1.408)	(1.851)	(2.044)	(2.048)	(2.055)
ln QG	0.872	0.910	0.937	0.937	0.938
	(3.904)	(3.347)	(3.619)	(3.759)	(3.781)
ln YG	0.125	0.113	0.104	0.104	0.103
	(1.961)	(1.668)	(1.613)	(1.630)	(1.624)
ln SC	−0.027	−0.331	−0.034	−0.034	−0.034
	(−0.560)	(−0.528)	(−0.590)	(−0.601)	(−0.603)
μ	—	—	−113.158	—	—
	—	—	(−0.003)	—	—
λ	—	1.629	16.460	0.824	0.803
σ_v	—	0.321	0.346	0.346	0.350
σ_u	—	0.522	5.701	0.285	0.281
σ	—	0.613	5.712	0.448	0.449
θ	—	—	—	3.512	3.300
P	—	—	—	—	0.858
Log-likelihood function	—	−37.6902	−37.126	−37.122	−37.085

* *t*-ratios are shown in parentheses.
** The coefficients are the same as those estimated by OLS.

of half-normal, truncated-normal, exponential and gamma distributions, respectively, and are all significant at the 1 per cent level (the critical value with 5 degrees of freedom is 15.086 at the 1 per cent level), thus leading to the rejection of the null hypothesis that the coefficients are equal to zero.

Without *a priori* justification of the assumptions imposed on the efficiency component, all four of the distributional assumptions; half-normal, truncated normal, exponential and gamma are examined still further. The estimation of terminal efficiency is conducted using the LIMDEP 8.0 econometric software package – the latest review of which can be found in McKenzie and Takaoka, 2003. Because estimation procedures yield merely the residuals ε rather than the inefficiency term u then, as suggested by Jondrow *et al.* (1982), the conditional expectation of u_k, conditioned on the realised value of the error term $\varepsilon_k = (v_k - u_k)$ is an estimator of u_k. Finally, the technical efficiency is calculated by $U_k = \exp(-u_k)$.

Under each of the four assumed possible distributional forms for the inefficiency term in the model, the efficiency of each terminal at different ports is reported in Appendix 2, with a summary of Appendix 2 provided in Table 5.4. The four distributional assumptions yielded different results for efficiency and an ANOVA yielded a value of $F = 9.18$, which is significant at the 1 per cent level (with a critical value of 3.86). The Spearman's rank order correlation coefficient between the efficiency rankings derived from the four assumed models ranges from 0.98 to 1, indicating a high level of consistency between the estimation procedures.

It is clear that most efficiency estimates generated by COLS are lower than those yielded by the four stochastic frontier models; a result that can be attributed to the imposition of various assumptions on its method of calculation. ANOVA reveals that $F = 95.76$ for the five sets of efficiency scores, which is significant at the 1 per cent level (with a critical value of 3.38). However, Spearman's rank order correlation coefficients between the efficiency rankings derived from COLS and each of the four models with predetermined distributional assumptions all exhibit a fairly high value of 0.99. Thus, the parametric estimation methods yield rather similar estimates of efficiency in terms of the ranking of firm-specific inefficiency. As a result, the performance of the deterministic models is qualitatively similar to that of the stochastic

Table 5.4 Summary of the efficiency yielded from alternative parametric approaches

	COLS	Half-normal	Truncated-normal	Exponential	Gamma
Mean	0.3831	0.6770	0.7650	0.7657	0.7842
Standard error	0.0217	0.0169	0.0155	0.0155	0.0156
Median	0.3759	0.7185	0.8106	0.8113	0.8298
Standard Deviation	0.1711	0.1334	0.1219	0.1219	0.1226
Sample Variance	0.0293	0.0178	0.0149	0.0149	0.0150
Kurtosis	2.3761	0.9169	5.4869	5.5429	5.9541
Skewness	1.2354	−0.9074	−1.9638	−1.9738	−2.0421
Range	0.9294	0.6640	0.6923	0.6929	0.7036
Minimum	0.0706	0.2292	0.2235	0.2231	0.2280
Maximum	1.0000	0.8932	0.9158	0.9160	0.9316
Count	62	62	62	62	62

frontier models. This finding strongly supports the argument raised by Ondrich and Ruggiero (2001) that the stochastic frontier model is not necessarily better than the deterministic frontier model.

5.2.3   Comparison between non-parametric and parametric frontier models using cross-sectional data

Figure 5.3 shows the average efficiency yielded by various non-parametric and parametric approaches. It is clear that COLS yields rather low average efficiency measures, while the FDH model yields much higher average efficiency scores. The rest of the estimation models yield reasonably high efficiency scores ranging from 0.57 to 0.78. ANOVA reveals that $F = 43.12$ for these eight sets of efficiency estimates, which is significant at the 1 per cent level (the critical value is 2.68).

Table 5.5 reports the Spearman's rank order correlation coefficient between the different levels of efficiency as estimated by the various model forms. The mostly high values of the coefficient (except for FDH *vis-à-vis* other alternative approaches) indicate that these alternative approaches yield rather similar estimates of efficiency in terms of the rankings of the firms under study.

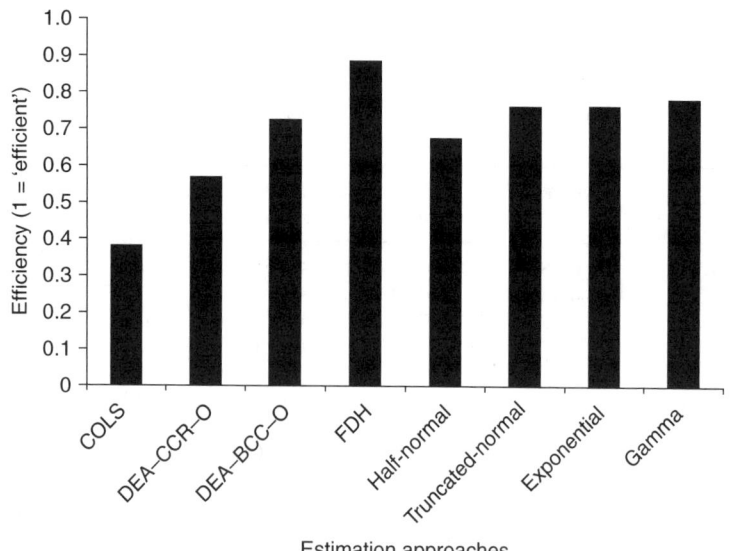

Figure 5.3   Comparison of efficiency yielded from various approaches.

Table 5.5 Spearman's rank order correlation coefficient

	DEA–CCR	DEA–BCC	Half-normal	Truncated-normal	Exponential	Gamma	COLS	FDH
DEA–CCR	1							
DEA–BCC	0.8123	1						
Half-normal	0.8049	0.6802	1					
Truncated-normal	0.7955	0.6724	0.9979	1				
Exponential	0.7955	0.6724	0.9979	1	1			
Gamma	0.7894	0.6708	0.9829	0.9812	0.9812	1		
COLS	0.8091	0.682	0.9975	0.9942	0.9942	0.9813	1	
FDH	0.6462	0.6349	0.4689	0.4704	0.4704	0.4617	0.4588	1

5.3 Panel data analysis

In this section, sample 2, that is, the panel data, are used as the basis for deriving the efficiency estimates of container port production. In a similar fashion to the testing that has been conducted on the cross-sectional data, a number of alternative approaches are used to examine the effects of model choice on the efficiency estimates that are produced.

5.3.1 DEA panel analysis

As with the analysis of cross-sectional data using DEA approaches, in the absence of categorical empirical *a priori* proof that the production function of container ports exhibits either constant or variable returns to scale, the DEA–CCR and DEA–BCC models were chosen from among several DEA models to analyse port production. Several alternative versions of DEA panel data analyses were implemented as part of this process. These included models that are integral to the Contemporaneous, Intertemporal and Window approaches to the estimation of efficiency using panel data. While it is relatively straightforward to calculate efficiency estimates using Contemporaneous and Intertemporal analyses, caution should be exercised in defining the window width for conducting a Window analysis. Ideally, it should be defined to correspond to the standard cycle time between technological innovations so that the efficiency estimates derived from Window analysis reflect solely the difference between the actual level of production of a firm and the best level of contemporaneous production. However, even if such a thing should exist, the technological innovation cycle time within the port industry is difficult to observe in practice and, therefore, in common with many previous studies of this kind, it is difficult to find a justification for the choice of window size. As such, the length of the window used herein is defined as three time periods in order to be consistent with the original work in Charnes *et al.* (1985). Six separate windows are represented as separate rows in Appendices 3 and 4. The average of the 18 DEA efficiency scores and their associated standard deviations are presented in the columns denoted 'Mean' and 'SD'. The efficiency estimates are reported in Appendices 3–8.

The approaches used in formulating Appendices 3–8 lend themselves to a study of 'trends' of efficiency over time. This is achievable through the adoption of a 'row view'. For instance, a cursory glance at Appendix 3 may prompt the inference that the efficiency of a container terminal differs significantly over time. Taking Hong Kong as an example, its efficiency varies from 0.7211 in 1992 to 0.8997 in 1999. Distinct from the other two approaches, Window analysis also lends itself to the examination of the 'stability' of efficiency within windows by the adoption of a 'column view'. For instance, by adopting this perspective, it is possible to observe that the efficiency of a firm within the different windows can also vary substantially. The observation of 'trend' and 'stability' in Window analysis reflects simultaneously both the absolute performance of a port over time and the relative performance of that port in comparison to the others in the sample.

Figure 5.4 depicts the development of the year-by-year average efficiency of all the container ports in the sample using Contemporaneous, Intertemporal and Window analysis, assuming in each case both the CCR and BCC model forms. It is clear from Figure 5.4 that the general trend of average efficiency for the results from applying Intertemporal analysis during the study period is upward, compared with the downward and almost flat trends (with some fluctuations) observed in the average efficiency estimates derived from applying Window and Contemporaneous analyses, respectively. The former can be explained by the fact that long-term technological advancement and managerial development provide an important impetus for improving productivity and efficiency. Within a shorter time period, defined by a window width of a mere three years in this research, the different firms (it is important to recognise that the same port observed at different time periods is treated as being different firms) are more likely to use the same or similar technology and management. In such a case, estimated efficiency results are not greatly influenced by the technology and management utilised. Figure 5.4 also shows that an advancement in technology does not necessarily imply an overall improvement in efficiency; a feature that can be shown by the

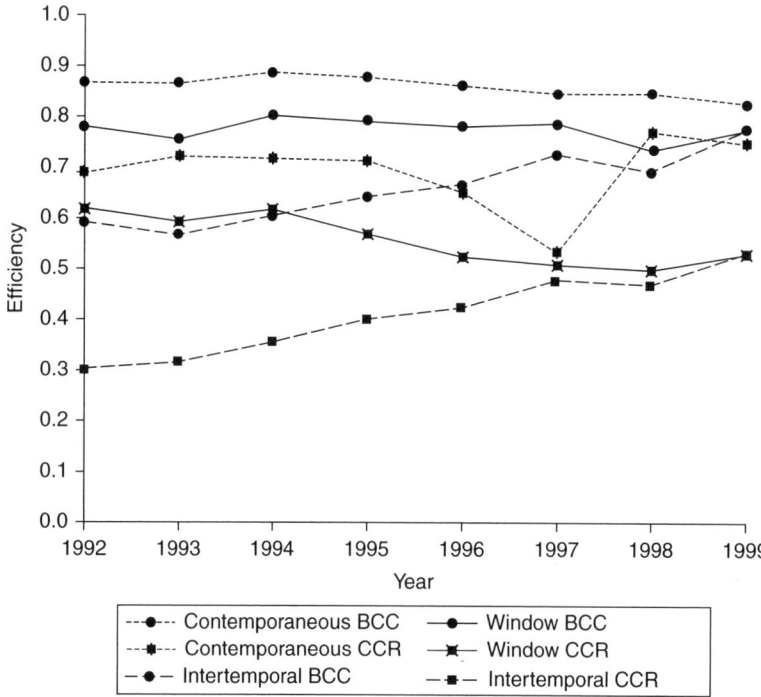

Figure 5.4   Year-by-year average efficiency for all container ports (1992–99)

decline in the average efficiency estimates from 1997 to 1998 when applying Intertemporal analysis.

Figure 5.4 also shows that the average efficiency for all of Contemporaneous, Window and Intertemporal analyses exhibits a downward trend. This is not surprising since it can be explained by the conceptual illustration shown in Figure 3.5 in Chapter 3. In this study, each port is compared with 29 other counterparts in the same set for a Contemporaneous analysis. However, each port is compared with 89 and 249 other counterparts for Window and Intertemporal analyses. A large sample is obviously more likely to make a firm appear inefficient. An ANOVA of the average efficiency of each port over time for Contemporaneous, Window and Intertemporal analyses ($F = 11.19$ and 9.96 corresponding, respectively, to DEA–CCR and DEA–BCC analyses) indicates that the means of the efficiency measures calculated using these three different approaches are significantly different at the 1 per cent level (with a critical value of 4.86). On the other hand, Spearman's rank order correlation coefficient between the efficiency derived from the three approaches ranges from 0.89 to 0.98. The high value of this coefficient indicates that the three approaches yield similar rankings of efficiency for container port production.

5.3.2   Time-invariant efficiency analysis using panel data

As discussed in Chapter 3, the time-invariant efficiency model is mainly divided into the fixed- and random-effects model. Within-group transformation and least squares dummy variable (LSDV) estimations are two equivalent methods of solving the fixed-effects models. The solution of random-effects models is, on the other hand, normally achieved through the application of any one of a generalised least squares (GLS) approach, the generalised least squares instrumental variable approach (GLS–IV, as suggested by Hausman and Taylor, 1981) and MLE.

These different approaches to estimation are based on different assumptions imposed on the model. It is of great importance to examine the validity of these assumptions within the context of the container port industry and, in particular, to the models suggested in this book. Note that the models suggested are based on the fact that only the most important facilities and infrastructure in container terminals are incorporated into the model as input variables. By adopting such a strategy, less important facilities such as mobile cranes and trucks are collectively reflected in the efficiency and random-effects u_{kt} and v_{kt}. When these less important facilities play a minor role in container yard production, it might be safe to assume that the input variables are uncorrelated with the efficiency and random-effects u_{kt} and v_{kt}. In this case, GLS and MLE are preferred because an assumption of uncorrelatedness underpins these estimation methods. On the other hand, if the ports in the sample rely heavily upon these less important facilities, which is particularly true of some small-scale container ports, it is dangerous to assume that efficiency and random effects (u_{kt} and v_{kt}) are uncorrelated with input variables. In this case, the within-group transformation and least squares estimation corresponding to the fixed-effects model, and the GLS–IV corresponding to the random-effects model, are the preferred estimation methods, as they are particularly adept at handling this situation.

With the exception of the GLS–IV estimation model, all of the other estimation routines mentioned above have been applied to the panel data in order to estimate the time-invariant efficiency measures. This was done in order to observe the robustness of the efficiency results in

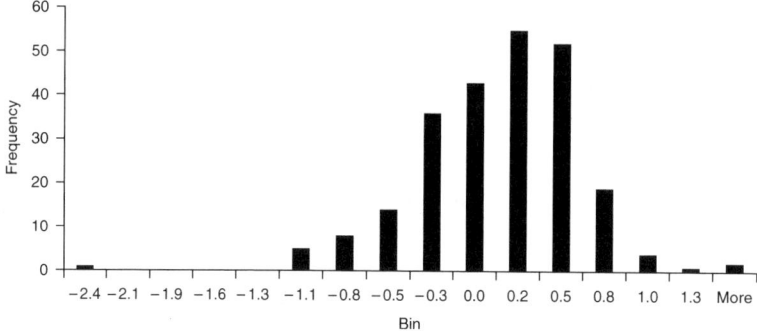

Figure 5.5   Skewness of the OLS residuals for panel data.

response to the various estimation methods applied. The GLS–IV is excluded not only because of its complexity, but also because of the characteristic consistency of efficiency estimates that are yielded by both GLS–IV and GLS (Hausman and Taylor, 1981). In other words, GLS efficiency estimation will, in principle, provide a reasonable approximation to GLS–IV estimates.

As far as MLE is concerned, as is always the case, the sign of the third moment and the skewness of the OLS residuals are determined from the sample data (Waldman, 1982) before the conclusion can be drawn that MLE can be appropriately conducted. In the case of the panel data, the skewness for the port production frontier model is −0.5662 and the histogram of the residuals, as shown in Figure 5.5, is clearly negatively skewed.

The Lagrange multiplier test statistic of 262.03 (see Breusch and Pagan, 1980) is markedly significant at the 1 per cent level (with a critical value of 6.63). This leads to the rejection of the null hypothesis that individual effects do not exist and to the acceptance of the fixed- and random-effects models.

Fixed and random effects generalised least squares are based on different assumptions and the estimated slopes are accordingly different. The Hausman test statistic of 17.85 (see Hausman, 1978) is marginally significant at the 1 per cent level (with a critical value of 15.09 and $df = 5$). A high Hausman test statistic implies that the null hypothesis that the individual effects are uncorrelated with the other regressors in the model should be rejected and that the fixed-effects model is more appropriate. However, the Hausman test statistic of 17.85 is rather close to the critical value, indicating that both the fixed- and random-effects models are equally suitable for use in deriving the efficiency value.

The estimated coefficients by alternative approaches and the time invariant efficiency of the container ports are reported in Table 5.6 and Appendix 9, respectively. The efficiency based on the assumption of an exponential distribution for the maximum likelihood estimation yields dramatically 'optimistic' results compared with the other three approaches. ANOVA reveals that $F = 73.66$, which is significant at the 1 per cent level (with a critical value of 3.96).

Table 5.7 reports the Spearman's rank order correlation coefficients, and shows that the efficiency results estimated by the MLE-exponential model is markedly different from the other three models. On the other hand, the other three models yield a similar pattern of efficiency. A revisiting of the ANOVA of these three models reveals that $F = 5.294$, which is similar to the critical value of 4.858 at the 1 per cent level. This leads to the conclusion that LSDV, GLS and MLE-half-normal models might be more appropriate than the MLE-exponential models.

Table 5.6   Frontier production function of container ports*

Parameters	OLS	Fixed effects within	Random effects		
			GLS	MLE-half-normal	MLE-exponential
Constant	10.6435	—	8.6757	9.5464	9.6645
	(25.6501)	—	(16.069)	(46.317)	(47.987)
ln TL	0.1901	0.8117	0.5968	0.5866	0.5544
	(2.2531)	(7.095)	(6.174)	(11.544)	(10.227)
ln TA	−0.0801	−0.6316	−0.1262	−0.1896	−0.1896
	(−1.8402)	(−0.563)	(−1.684)	(−2.002)	(−2.180)
ln QG	0.6554	0.3848	0.3783	0.4231	0.4262
	(6.7910)	(3.011)	(3.514)	(5.413)	(5.628)
ln YG	0.1826	−0.0067	0.1181	0.1195	0.1375
	(5.9887)	(−0.107)	(2.681)	(2.877)	(3.559)
ln SC	0.0147	−0.0095	−0.094	0.0632	0.0121
	(0.6501)	(0.278)	(−0.338)	(0.156)	(0.326)
λ	—	—	—	2.592	1.941
σ_v	—	—	—	0.787	0.306
σ_u	—	—	—	0.844	0.594
Log-likelihood function	—	—	—	−96.96	−99.51

* Standard errors are shown in parentheses.

Table 5.7 Spearman's rank order correlation coefficient

	Within	GLS	Half-normal	Exponential
Within	1			
GLS	0.7281	1		
Half-normal	0.8438	0.8803	1	
Exponential	−0.8202	−0.9172	−0.9969	1

Table 5.8 Frontier production function of container ports*

Parameters	OLS	MLE-half-normal
Constant	10.6435	—
	(25.6501)	—
ln TL	0.1901	−0.0188
	(2.2531)	(−0494)
ln TA	−0.0801	−0.091
	(−1.8402)	(−0.419)
ln QG	0.6554	0.7836
	(6.7910)	(24.156)
ln YG	0.1826	0.1653
	(5.9887)	(11.702)
ln SC	0.0147	−0.01
	(0.6501)	(0.999)
λ	—	2.131
σ_v	—	0.3100
σ_u	—	0.6606
Log-likelihood function	—	−92.92

* Standard errors are shown in parentheses.

5.3.3   Time-varying efficiency analysis using panel data

The stochastic frontier model in a true fixed-effects formulation according to Greene (2002) is chosen for application in this research. The structural coefficients and time-varying efficiency estimated by MLE are reported in Table 5.8 and Appendix 10, respectively. Apart from the coefficient for ln TL, the other estimated coefficients are similar to those estimated by OLS. This is because both OLS and MLE yield consistent estimates (MLE is asymptotically consistent).

Because MLE is the only approach utilised to estimate the time-varying efficiency, it is hard to evaluate its validity. A feasible approach is to compare the average efficiency of each port over time with the time-invariant efficiency estimated from the LSDV, GLS and MLE-half normal estimation results. An ANOVA shows that $F = 23.16$, which is significant at the 1 per cent level (with a critical value of 3.955). Spearman's rank order correlation coefficients between LSDV and the average time-varying efficiency, between GLS and the average time-varying efficiency, and between the MLE-half normal and average time-varying efficiency are -0.16, 0.07 and -0.06, respectively. This indicates that MLE yields a different time-varying efficiency from its time-invariant counterpart.

5.3.4   Comparison between non-parametric and parametric frontier models using panel data

Figure 5.6 shows a comparison of average efficiency estimates derived from the alternative non-parametric and parametric approaches. As far as the efficiency results generated by the parametric approaches are concerned, average time-varying efficiency is higher than for the time-invariant approaches. Most of the efficiency scores generated by the non-parametric approaches are higher than those using parametric approaches. The different trends in efficiency estimates over time that are shown in Figure 5.6 clearly demonstrate that the choice of estimation technique, model specification and the assumptions imposed on each model clearly do matter where efficiency estimation is concerned.

Figures 5.6 shows that the average efficiency across firms estimated by the Intertemporal CCR and BCC models and the time-varying parametric models demonstrate a similar upward trend. This result is also in accordance with expectations, not only because technological innovation is always the impetus for improvements in efficiency, but also because the long life-cycle characteristics of equipment and infrastructure that are deployed in the container port industry require that the initial planning and investment are done on the basis of long-term future planning that usually has growth in business as a fundamental premise. In other words, the assumption is made that design capacity, although originally underutilised, can gradually be mobilised and the efficiency of port production can be enhanced as time passes. On the other hand, when the average efficiency estimates of individual ports that are yielded by these three alternative approaches is compared, ANOVA (with $F = 22.20$ and a critical value of 4.86 at the 1 per cent level) and Spearman's rank order correlation coefficient (0.48 between the Intertemporal CCR and BCC models efficiency estimates, -0.10 between the Intertemporal CCR and time-varying parametric model estimates, 0.19 between the Intertemporal BCC and time-varying parametric

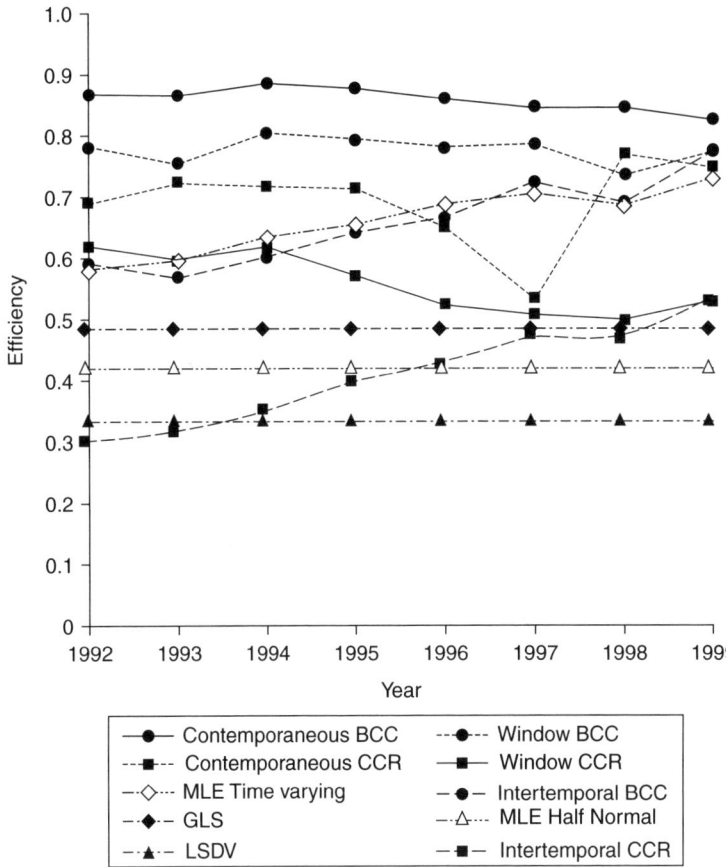

Figure 5.6 Comparison of average efficiency by alternative approaches.

model estimates) reveal that the efficiency estimated by different estimation models is significantly different from each other. This result, where the average efficiency level is similar across firms but different for individual ports over time, can be explained by the properties of parametric production frontier estimation: the estimated coefficients representing the movement of average efficiency across firms is unconditionally unbiased and consistent. When the efficiency of individual firms is conditionally estimated, the assumptions imposed may be too strong or even erroneous. Similar empirical proof was found in hospital production by Banker *et al.* (1986), who found that the parametric model may, in fact, simply be 'averaging' diametrically opposite behaviour.

To observe the trend in the efficiency estimates of individual ports that are derived from alternative approaches, as shown in Figure 5.7, the container port of Hong Kong has been selected. The efficiency of the container port of Hong Kong, as estimated by most approaches, demonstrates rather high levels and accords with the general assumption that, as the world's leading port in terms of container throughput, it has been able to keep abreast of both the development of state-of-the-art technology and sophisticated management. Thus, it is not only determined to be, but has also proved itself to be, an efficient entity.

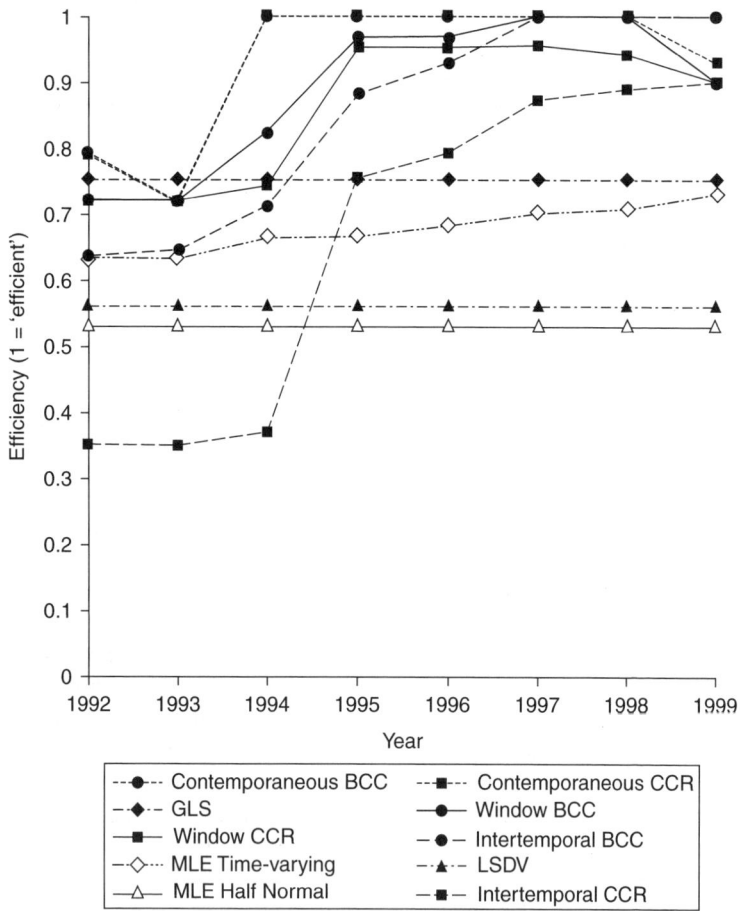

Figure 5.7　Comparison of the efficiency of Hong Kong container port yielded by alternative approaches.

5.4 Summary

The importance of obtaining accurate measures of efficiency cannot be emphasised enough in order that the results can then be used to analyse the nature of production. Existing theory does not prescribe which of the vast array of alternative approaches is superior to others, but claims that every approach has its own strengths and weaknesses.

This chapter contributes to existing research in that a comprehensive comparison of alternative approaches to estimating efficiency is conducted. In so doing, an empirical contribution to the debate over the relationship between efficiency results and the choice of estimation model has been made.

Based on the theoretical discussion in Chapter 3 relating to alternative measures of efficiency, this chapter sets out to compare estimates of efficiency using cross-sectional data. The non-parametric and parametric models applied in this section included DEA, FDH, COLS and MLE. As far as non-parametric estimation approaches are concerned, this chapter contributes to the existing literature on the topic by providing a comparative study of the container port industry using the two non-parametric approaches of DEA and FDH. A detailed analysis of the efficiency estimates yielded by the two DEA models (CCR and BCC), together with the FDH model, confirms that the DEA and FDH mathematical programming methodologies tend to give significantly different results. In this aspect, it should be emphasised that the FDH model identifies more efficient container ports than inefficient ones. This contrasts with the DEA model, which identifies more inefficient ports than efficient ports. This leads to the conclusion that the sample size used in the analysis is not sufficiently large to develop a meaningful facet or frontier in FDH models and that the DEA model is more informative and appropriate for any further analysis.

ANOVA and Spearman's rank order correlation coefficient revealed that, although the efficiencies estimated by the alternative approaches differ significantly from each other, the ranking of the sample ports using alternative approaches is rather similar. This leads to the conclusion that the efficiency estimated by alternative approaches exhibits a similar pattern of efficiency distribution.

The chapter proceeded to compare the efficiency estimates derived from alternative non-parametric and parametric models using panel data. Compared with estimates of efficiency using cross-sectional data, it is much more difficult to estimate efficiency using panel data. In the context of non-parametric models, the DEA Contemporaneous,

Intertemporal and Window approaches were used to derive the efficiency of container port production. In essence, both Contemporaneous and Intertemporal models constitute merely extreme forms of Window analysis when the window width equals one and the whole study period T, respectively. It is of great importance to distinguish between Contemporaneous, Intertemporal and Window analyses. While Contemporaneous analysis examines whether a firm is able to follow the best industry practice during the same production period, Intertemporal analysis is able to measure whether a firm can simultaneously both keep up with best practice and keep abreast of technological innovations. By adjusting the width of the window in Window analysis, it is possible to observe the movement of the production frontier over time.

The estimation of efficiency by parametric models using panel data was then discussed on the basis that the efficiency of container port production may be either time-invariant or time-varying. It was found that most estimates of the efficiency of individual ports using these parametric approaches differ greatly from those using DEA approaches. However, the average efficiency of container ports across firms due to the time-varying analysis of panel data is similar to DEA Intertemporal analysis. This leads to the conclusion that parametric approaches are more capable of discovering the average efficiency across firms than of estimating the efficiency of individual ports. This can be attributed to the inherent weaknesses of the strong assumptions that most parametric estimation models rely upon.

This chapter finds that the parametric estimation approach also suffers from an inadequacy in terms of methodological flexibility. This is particularly true for time-varying estimates of efficiency with panel data. Several established approaches in this respect were discovered to be extremely difficult to apply for a variety of reasons. In terms of the capability of models to reflect real production, non-parametric approaches have been found to have an advantage, particularly since, compared to the family of parametric approaches, they have far fewer strong assumptions imposed on the efficiency estimates that are derived from them. In sharp contrast, there are numerous potential difficulties when parametric approaches are applied to efficiency estimation. These include having to pre-define the production function and to make assumptions on the distributions of errors. For instance, the Cobb–Douglas model is used in this book to represent the production function. However, its suitability in any empirical application, including that contained within this book, may be called into question. At the same time, however, the choice of alternative functional forms in this respect is rather limited.

To sum up, the empirical results contained in this book lead us to advocate the use of various non-parametric estimation approaches, particularly various DEA approaches. It is this premise that constitutes the basis for a further analysis of the economic theory underpinning container port production that is presented in Chapter 6.

6

A Theory of Container Port Production and its Empirical Validation

6.1 Introduction

This chapter mainly serves to test the hypotheses formulated in Chapter 2 by using the efficiency results derived in Chapter 5. By so doing, light can be shed on the relevance of fundamental economic theory in underpinning the operation of the container port industry. Despite their respective strengths and weaknesses, as concluded from the analysis presented in Chapter 5, non-parametric and parametric models for analysing efficiency either generate similar estimates of efficiency when utilising cross-sectional data, or the former yield more convincing estimates of efficiency than the latter when significant differences exist between the efficiency scores estimated by the two approaches. For these reasons, in this chapter, the results derived from the non-parametric set of models will constitute the fundamental basis for testing the hypotheses expounded in Chapter 2.

This chapter is structured as follows. In Section 6.2, the six hypotheses formulated in Chapter 2 are tested. During this process, the relevant economic theories that are deemed to theoretically underpin the structure and conduct of the container port industry are analysed. Section 6.3 summarises and concludes this chapter.

6.2 Testing the hypotheses

Hypothesis 1: The technical efficiency of container ports has improved with time.

This hypothesis can be properly tested by tracking the year-on-year average efficiency of all of the container ports in the sample using the Contemporaneous, Intertemporal and Window analyses reported in Appendices 3–8 and summarised in Figure 5.4. As discussed in Chapter 5, the general upward trend in the average efficiency estimates under Intertemporal analysis indicates that the overall efficiency of the container port industry has improved with time, at least during the period covered by the sample. This is due to both long-term technological advancement and to the development and enhancement of managerial acumen. However, the downward and almost flat trends respectively observed in the average efficiency estimates derived from Window analysis and from Contemporaneous analysis indicate that not all ports have been able to follow best practice during the sample period.

Thus, the decision as to whether to accept or reject hypothesis 1 hinges on the definition of efficiency that is applied. Hypothesis 1 cannot be rejected if the efficiency under study refers to an overall efficiency that is influenced by technological innovation and management. However, hypothesis 1 can and should be rejected if it is held to refer solely to whether a firm follows best practice at any particular time.

Hypothesis 2: Efficiency increases as the scale of a container port increases. In other words, a large-scale container port is more likely to be associated with high efficiency than a small one.

To test hypothesis 2, efficiency estimates based on both cross-sectional and panel data are utilised. This is because in the case of the former every single firm is observed only once and, hence, the efficiency estimates that result may be influenced by random effects and, therefore, may be misleading. This potential drawback is largely overcome through the use of panel data. Another advantage of using panel data is that the sample size increases from 62 to 240 and, in consequence, the statistical validity of the results and inferences drawn from the analysis of this enlarged sample are much more likely to provide more reliable conclusions than would otherwise be the case.

Table 6.1 shows the relationship between scale of production and efficiency, as measured by the Pearson correlation coefficient (ranging from 0.11 to 0.32) and Spearman's rank order correlation coefficient (ranging from 0.03 to 0.35). The fact that the signs for both the Pearson correlation coefficient and Spearman's rank order correlation coefficient are positive does imply that the production volumes of container ports and terminals are indeed positively associated with efficiency scores. On the

Table 6.1　Relationship between production size and efficiency

Data nature	Estimation model	Container throughput versus efficiency	
		Pearson correlation	Spearman's rank order correlation
Cross-sectional	DEA–CCR–O	0.2903	0.3474
	DEA–BCC–O	0.3193	0.3232
	FDH	0.1147	0.0662
Panel	DEA–CCR Window	0.1815	0.0577
	DEA–BCC Window	0.2718	0.1753
	DEA–CCR Intertemporal	0.2842	0.1651
	DEA–BCC Intertemporal	0.3161	0.2500
	DEA–CCR Contemporaneous	0.1795	0.0620
	DEA–BCC Contemporaneous	0.1659	0.0349

other hand, the small absolute values of the Pearson correlation coefficient and Spearman's rank order correlation coefficient would seem to indicate that the efficiency of a port is not significantly related to scale of production. This differs from the usual informal supposition that prompted the formulation of hypothesis 2 in the first place and also contradicts some prior empirical evidence (e.g. Kim and Sachish, 1986; Tongzon, 1993; Jara-Diaz *et al.*, 1997; Drewry Shipping Consultants, 1998; Jara-Diaz *et al.*, 2002; Robinson, 2002).

One possible explanation for this is that the apparent relative inefficiency of large container ports is due, not to what may seem to be managerial shortcomings, but rather to the overcapacity that results from the more intensive efforts of larger ports to maintain or enhance productivity levels. These efforts manifest themselves through the injection of more significant investment funds than their smaller counterparts and, therefore, the wider availability of larger numbers of, and more sophisticated, items of equipment. In so doing, the competitiveness of larger container ports or terminals is thereby increased, relative to the rest of the market. In summary, although empirical support for hypothesis 2 using this methodology is hardly categorical, neither can it be rejected with alacrity. Instead, it is most appropriate to surmise that it should be accepted but with caveats.

Hypothesis 3: *Fluctuations in the technical efficiency of container ports are not related to scale of production.*

Panel data provide the basis upon which hypothesis 3 is tested. To examine the degree to which efficiency fluctuates with scale of production, the standard deviation of efficiency estimates of each port over time and the mean container throughput at each port over time are correlated using the Pearson correlation coefficient and Spearman's rank order correlation coefficient. The results of this analysis are reported in Table 6.2 where the calculation of the standard deviation of efficiency estimates derived from applying Window analysis is based on all of the efficiency estimates for an individual container port in different windows.

The small absolute values of the Pearson correlation coefficient and of Spearman's rank order correlation coefficient in Table 6.2 show that only a weak correlation exists between scale of production and the standard deviation of a container port's efficiency estimates over time (as a proxy for fluctuations in efficiency). Since these correlations are statistically insignificant, in a practical sense, this implies that the efficiencies of all of the ports in the sample exhibit a similar level of fluctuation over time. As such, hypothesis 3 cannot be rejected.

Hypothesis 4: *Container ports faced with external competition are more efficient than their counterparts in a less competitive market.*

The first step in testing hypothesis 4 is to develop a quantitative measure of the level of inter-port competition. Among the numerous methods that have been proposed for doing this (see Alam, 1984; Amato, 1995; Riccardo, 2000; Nanuenberg *et al.*, 2001), the Herfindahl–Hirschman index (better known as the Herfindahl index) has been selected as the

Table 6.2 Relationship between scale of production and fluctuations in efficiency

Data nature	Estimation model	Container throughput versus efficiency fluctuations	
		Pearson correlation	*Spearman's rank order correlation*
Panel	DEA–CCR Window	−0.1272	−0.1373
	DEA–BCC Window	−0.4014	−0.5150
	DEA–CCR Intertemporal	0.0173	−0.0202
	DEA–BCC Intertemporal	−0.1432	−0.0889
	DEA–CCR Contemporaneous	−0.1439	−0.0883
	DEA–BCC Contemporaneous	−0.2962	−0.1405

most appropriate measure of the level of competition for the purposes of this study. Although strictly speaking this is a measure of the degree of concentration (rather than competition) within an industry, classical economic theory (as well as numerous worldwide investigations into anti-competitive behaviour) would imply that industry concentration has a direct bearing upon the degree of competition in the market. Later refinements, such as that of the competitive markets hypothesis due to Baumol *et al.* (1982) and concepts such as the difference between 'competition *for* the market' and 'competition *in* the market' would seem to refute the necessarily direct relationship between market concentration and the degree of competition faced by incumbents. Nevertheless, because of its numerous advantages and its suitability for application to the container port industry, the Herfindahl–Hirschman index (HHI) is the preferred metric for use within this study. Rhoades (1993) presents an excellent example to show the practical importance of the HHI as follows:

> To facilitate and simplify the applications of the antitrust laws regarding mergers, in 1982 the Department of Justice published formal numerical guidelines for horizontal mergers (those between firms operating in the same product and geographic markets) based on the Herfindahl index (HHI).

The HHI accounts for the number of firms in a market, as well as their concentration, by incorporating the relative size (measured by market share) of all firms in a market. It is calculated by squaring the market shares of all firms in a market and then summing the squares, as shown in equation (6.1)

$$\text{HHI} = \sum_{k=1}^{K} (\text{MS}_k)^2 \tag{6.1}$$

where MS_k represents the market share of firm k, and there are K firms in the same market. In the case of the container port industry, the firms refer to competitive container ports serving the same or an overlapping hinterland.

Closer investigation of equation (6.1) reveals that, as a result of squaring the market shares of individual firms in the market, the HHI gives a much heavier weighting to firms with a large market share than to firms with a small share. This feature of the HHI corresponds to the theoretical notion in economics that the greater the concentration of output in a small number of firms (corresponding to a high HHI in equation (6.1)), the greater the likelihood that, *ceteris paribus*, competition in the market will be weak. In contrast, if concentration is low, reflecting a large number of firms with small market shares (corresponding to a low HHI in equation (6.1)), competition will tend to be quite vigorous. The HHI reaches a maximum value of 10,000; when a monopoly exists in which one firm has 100 per cent of the market, then HHI = 10,000. By contrast, the HHI takes on a very small value, theoretically approaching zero, in a purely competitive market in which there are many firms with small market shares. For instance, in a market with 100 firms where each has a 1 per cent share of the market, HHI = 100.

To sum up, the HHI is used to measure the concentration of market shares. As previously mentioned, in measuring competition, the assumption behind the HHI is that a low level of concentration is expected to be accompanied by a high degree of competition and vice versa. This assumption is particularly true for inter-port competition in the container port industry. Figure 6.1 shows that the overall level of competition in the container port industry as measured by HHI demonstrates an increasing trend over time (i.e. a decreasing value of the HHI over time indicates an increase in the level of competitive intensity). This result accords with the widely accepted view of the general trend in the container port industry over recent years, as referred to in previous important studies of inter-port competition that have included, among others, Notteboom and Winkelmans (2001), Heaver *et al.* (2001), Meersman and van de Voorde (2002) and Cullinane *et al.* (2004).

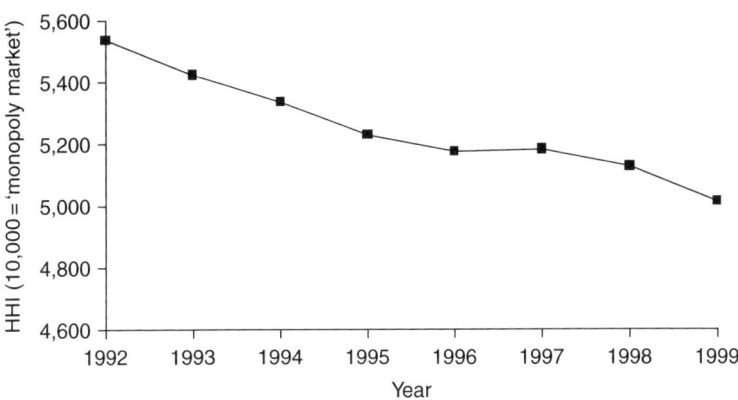

Figure 6.1 Overall level of concentration in the container port industry as measured by the HHI.

In principle, hypothesis 4 can be tested through analyses based on either cross-sectional or panel data. The former is conducted by comparing a port in an inter-port competitive market with its counterparts in a less competitive market, while the latter is undertaken by observing movements in the value of efficiency estimates over time and across firms. However, as mentioned earlier, some drawbacks always exist with cross-sectional data analysis because each firm is observed only once and, therefore, random effects may be present and might be exerting a disproportionate influence on the results. To avoid this potential problem, the panel data are further analysed using both DEA Intertemporal and Contemporaneous models.

The geographic locations of most of the container ports in the sample are shown in Figure 6.2. The proximity of some of the port locations roughly define which ports compete in the same market. For instance, the most important hub ports in Europe compete directly with each other. As an integral element in this part of the ongoing analysis, the hinterland that each port serves has been extensively investigated in order to ensure that sets of inter-competitive ports that share an overlapping hinterland are properly defined.

The market definition and the HHI for each port in the market are reported in Table 6.3. If hypothesis 4 is true, the relationship between the HHI and the efficiency of individual ports should resemble Figure 6.3. The relationship between inter-port competition and the efficiency estimates derived from the DEA–CCR, DEA–BCC and FDH models are plotted in Figures 6.4–6.7. It is clear from these figures that there is no clear-cut relationship between the HHI and the efficiency of container ports. Further evidence for this conclusion is provided by a low Pearson correlation coefficient and low Spearman's rank order correlation coefficient between the inter-port competition index (as proxied by the HHI) and the efficiency estimates that have been derived by the application of the different techniques, as reported in Table 6.4. On the basis of both cross-sectional and panel data analyses, therefore, hypothesis 4 is rejected. The positive signs of the estimates do, however, imply that a less competitive environment leads, to some extent, to higher efficiency. As highlighted by van den Broeck *et al.* (1994, p. 274)

> From an economic point of view, the need to survive in a competitive environment of most economic units induces a belief that many of them are close to the frontier, i.e., full efficiency. However, given the dynamic character of competition itself, strategic policies in the long run (secular inefficiency) could keep units away from their frontier. In many cases, this will be compounded with organisatorial [*sic*] inefficiency in the short run.

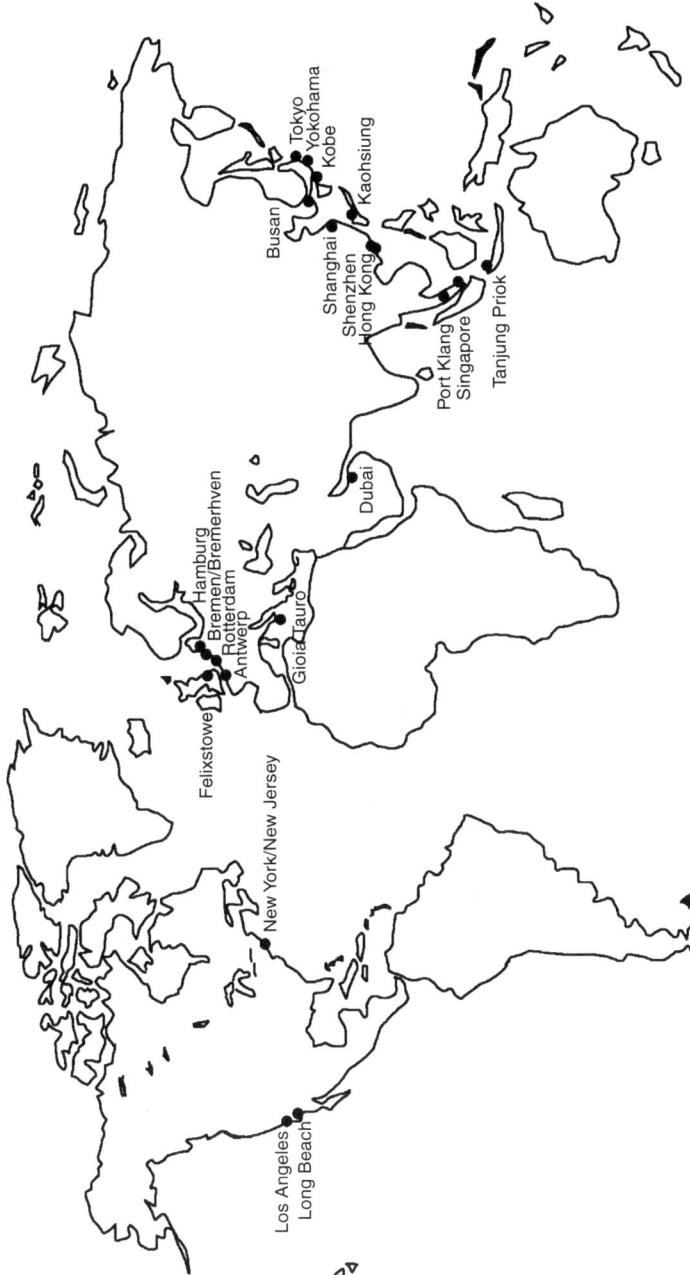

Figure 6.2 Geographic locations of the world's top container ports.

124

Table 6.3 Markets and inter-port competition

Market	Terminal	Port	Country/nation	HHI
Pearl River	HIT	Hong Kong	China	2,409
Delta	MTL	Hong Kong	China	
	Terminal 3	Hong Kong	China	
	Cosco-HIT	Hong Kong	China	
	Huangpu	Huangpu	China	
	Yantian	Shenzhen	China	
	Shekou	Shenzhen	China	
	Chiwan	Shenzhen	China	
Central	Shanghai	Shanghai	China	7,338
China	Nanjing	Nanjing	China	
	Ningbo	Ningbo	China	
Northern	Dalian	Dalian	China	3,596
China	Qingdao	Qingdao	China	
	Tianjin	Tianjin	China	
Southern	Xiamen	Xiamen	China	6,022
China	Fuzhou	Fuzhou	China	
Busan	Jasungdae	Busan	South Korea	1,937
	Shinsundae	Busan	South Korea	
	Uam	Busan	South Korea	
	Gamman G	Busan	South Korea	
	Gamman Hanjin	Busan	South Korea	
	Gamman Hyundai	Busan	South Korea	
	Gamman K	Busan	South Korea	
Japan	Kinjo Pier	Nagoya	Japan	2,760
	NCB	Nagoya	Japan	
	Tokyo	Tokyo	Japan	
	Kobe	Kobe	Japan	
	Yokohama	Yokohama	Japan	
Taiwan	Keelung	Keelung	Taiwan	6,583
	Kaohsiung	Kaohsiung	Taiwan	
Southeast	Singapore	Singapore	Singapore	5,736
Asian	Klang Container	Port Klang	Malaysia	
countries	Klang Port	Port Klang	Malaysia	
	Tanjung Priok	Tanjung Priok	Malaysia	
	South Habour	Manila	Philippine	
	Manila International	Manila	Philippine	
West	Rotterdam	Rotterdam	Netherlands	1,484
Europe	Burchardkai	Hamburg	Germany	
	Eurokai	Hamburg	Germany	
	TCT Tollerort	Hamburg	Germany	
	Unikai	Hamburg	Germany	

Continued

Table 6.3 Continued

Market	Terminal	Port	Country/nation	HHI
	Gioia Tauro	Gioia Tauro	Italy	
	Algeciras	Algeciras	Spain	
	Europe Terminal	Antwerp	Belgium	
	Seaport	Antwerp	Belgium	
	Noord Natie	Antwerp	Belgium	
	Noordzee	Antwerp	Belgium	
	Felixstowe	Felixstowe	UK	
	Bremen/ Bremerhaven	Bremen/ Bremerhaven	Germany	
North America	Los Angeles	Los Angeles	USA	3,432
	Long Beach	Long Beach	USA	
	New York/ New Jersey	New York/ New Jersey	USA	
Thailand	Laem Chabang	Laem Chabang	Thailand	5,350
	Bangkok	Bangkok	Thailand	
Middle East	Dubai	Dubai	UAE	10,000
Sri Lanka	Colombo	Colombo	Sri Lanka	10,000

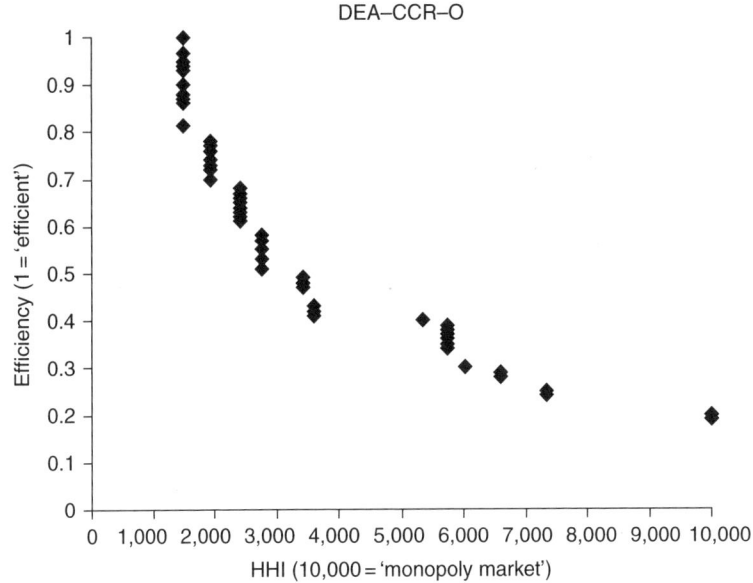

Figure 6.3 Hypothesised relationship between inter-port competition and efficiency.

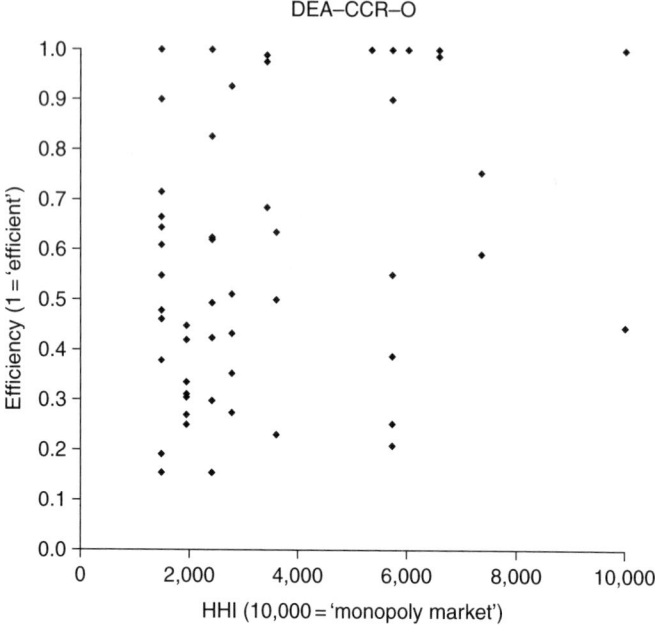

Figure 6.4 Relationship between inter-port competition and efficiency with reference to cross-sectional data (DEA–CCR–O).

Hypothesis 5: *The efficiency of container ports improves as ownership moves towards greater private sector participation.*

Song *et al.* (2001) examined the ownership status of some of the world's leading container ports in 1996. As was shown in Table 2.4, the ownership of the ports in their study were classified according to the port function matrix developed by Baird (1995, 1997). Their work constitutes the starting point for this research. However, the ownership of some ports has changed in recent years. For example, several ports in China have transferred their ownership from 'PUBLIC' to 'PUBLIC/private'. Even the world-renowned PUBLIC port of Singapore has also made some significant changes to its administrative and organisational structure, as detailed in Cullinane and Song (2001). As such, some amendments are necessary. Within the study period (1992–99), the main changes have taken place in several Chinese container ports, as summarised in Table 6.5. The final classification of the sample container ports is summarised in Figure 6.8.

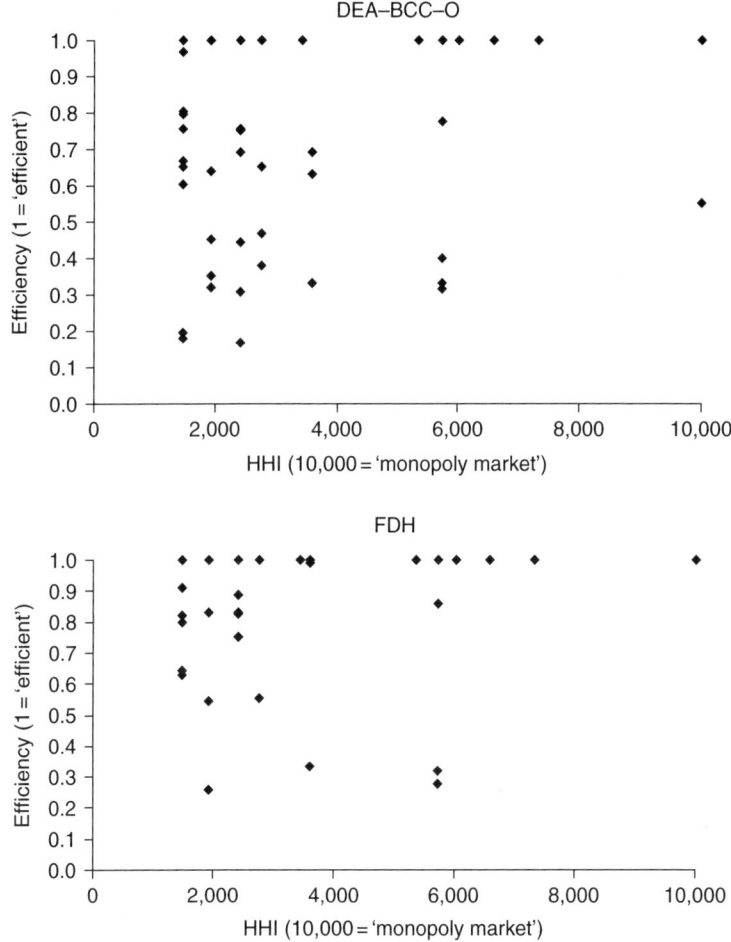

Figure 6.5 Relationship between inter-port competition and efficiency with reference to cross-sectional data (DEA–BCC–O, FDH).

Panel data and both DEA Contemporaneous and Intertemporal analyses have been selected to properly reflect the structural change in port ownership over time. Figures 6.9 and 6.10 depict the distribution of the relationship between port ownership and efficiency as respectively estimated by the DEA–CCR and BCC Contemporaneous models. Based on the samples analysed, the period covered by the analysis and the methods used in the analysis, it is clear that on average, 'PUBLIC' and

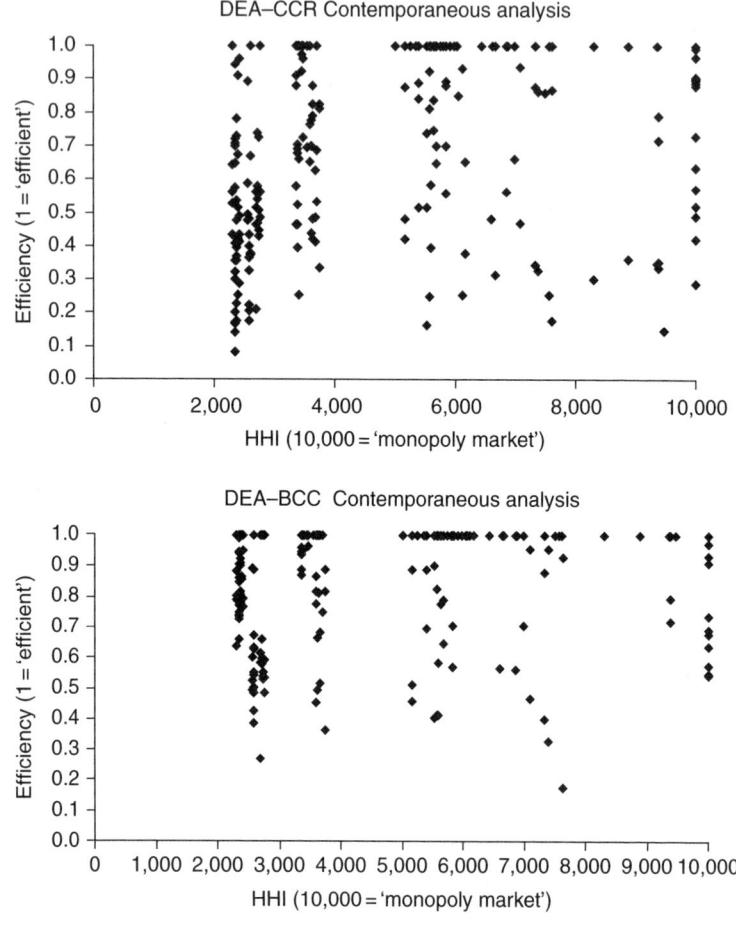

Figure 6.6 Relationship between inter-port competition and efficiency with reference to panel data (Contemporaneous analysis).

'private/PUBLIC' ports perform better than 'PUBLIC/private' and 'private' ports. As such, hypothesis 5 does not hold true according to a comparison between the efficiency of 'PUBLIC' and 'PUBLIC/private' ports. However, it should be emphasised that because 'private/PUBLIC' and 'private' ports each only comprise eight observations, their lack of representation in the sample is likely to lead to some bias in the results achieved. On the other hand, 'PUBLIC' and 'PUBLIC/private' ports comprise 44 and 180 observations respectively. Clearly these relatively

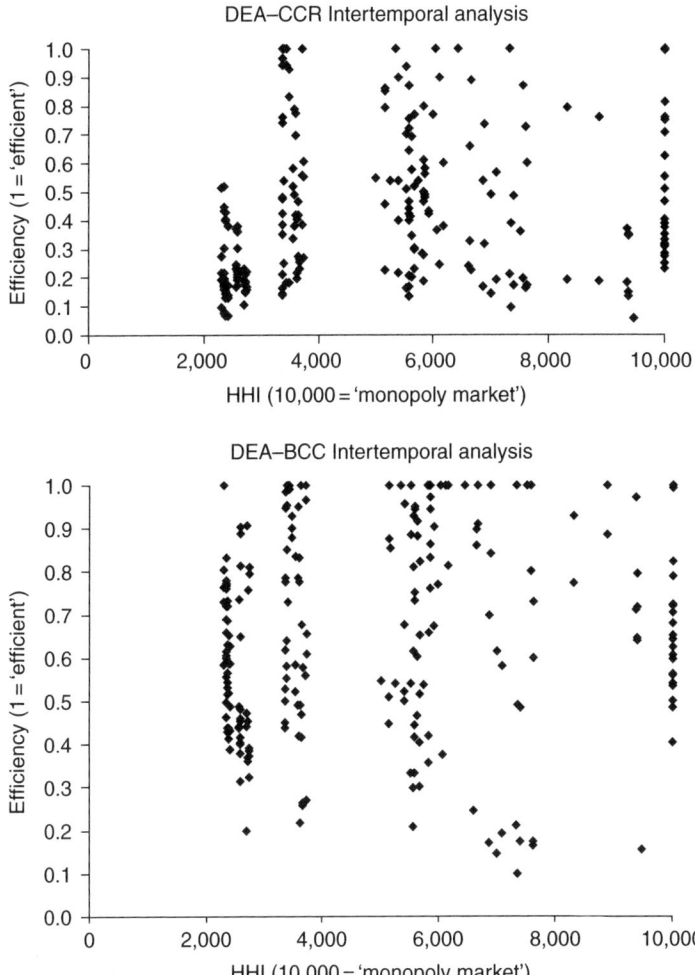

Figure 6.7 Relationship between inter-port competition and efficiency with reference to panel data (Intertemporal analysis).

large and significant representations within the sample, mean that any comparison of relative efficiency between these two categories of ports is likely to be much more convincing.

During the period over which the sample data was collected, the ownership of three container ports in China (Shanghai, Xiamen and

Table 6.4 Relationship between inter-port competition and efficiency

Data nature	Estimation model	HHI versus efficiency	
		Pearson correlation	Spearman's rank order correlation
Cross-sectional	DEA–CCR–O	0.2870	0.2193
	DEA–BCC–O	0.1887	0.1348
	FDH	0.0656	0.1059
Panel	DEA–CCR Contemporaneous	0.3082	0.3702
	DEA–BCC Contemporaneous	0.1674	0.2069
	DEA–CCR Intertemporal	0.2634	0.3847
	DEA–BCC Intertemporal	0.1352	0.1579

Table 6.5 Ownership status of the sample container ports

	Port	PUBLIC	PUBLIC/ private	PRIVATE/ public	PRIVATE
1	Hong Kong			√	
2	Singapore	√			
3	Busan		√		
4	Kaohsiung		√		
5	Shanghai*	√	√√		
6	Rotterdam		√		
7	Los Angeles		√		
8	Hamburg		√		
9	Long Beach		√		
10	Antwerp		√		
11	Port Klang		√		
12	Dubai		√		
13	New York/New Jersey		√		
14	Bremen/Bremerhaven		√		
15	Felixstowe				√
16	Manila		√		
17	Tokyo		√		
18	Qingdao	√			
19	Yokohama		√		
20	Laem Chabang		√		
21	Tanjung Priok		√		
22	Kobe		√		
23	Nagoya		√		
24	Keelung		√		
25	Colombo		√		
26	Dalian**	√	√√		

Continued

Table 6.5 Continued

	Port	PUBLIC	PUBLIC/ private	PRIVATE/ public	PRIVATE
27	Huangpu	√			
28	Nanjing		√		
29	Tianjin	√			
30	Xiamen***	√	√√		

* The largest joint venture terminal operation is Shanghai Container Terminal. It was established in 1993 between Shanghai Port Authority and Hong Kong-based Hutchison Whampoa Group (source: http://www.tdctrade.com/shippers/vol23_3/vol23_3_ports07.html). Thus, the port of Shanghai was a PUBLIC port during the period 1992–93 and a PUBLIC/private port from 1994 to 1999.

** Dalian container terminal was set up in July 1996 between the Port of Singapore Authority and Dalian port authority (source: http://www.dct.com.cn/). Thus, Dalian port was a PUBLIC port during the period 1992–96 and a PUBLIC/private port between 1997and 1999.

*** Xiamen International Container Terminal has been in operation since 1997 (http://www.hph.com.hk/business/ports/china/delta ports/xiamen.htm). Thus, Xiamen port was a PUBLIC port during the period 1992–96 and a PUBLIC/private port from 1997 to 1999.

Source: Derived from Song *et al.* (2001), Cullinane and Song (1998, 2001) and various official sources from the Chinese Mainland (in Chinese).

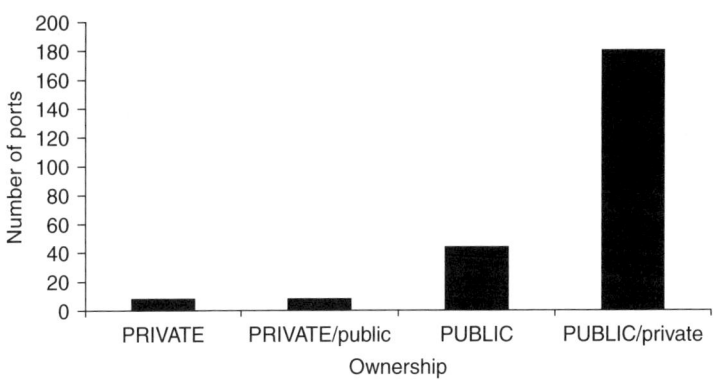

Figure 6.8 Ownership classification of the container ports in the sample.

Dalian) was transferred from 'PUBLIC' to 'PUBLIC/private'. It is interesting to look in detail at the change in the efficiency of these ports before and after their change in ownership. This can be seen in Figures 6.11 and 6.12. The dashed and solid lines in the figures denote, respectively, the level of estimated efficiency of the two container ports before and

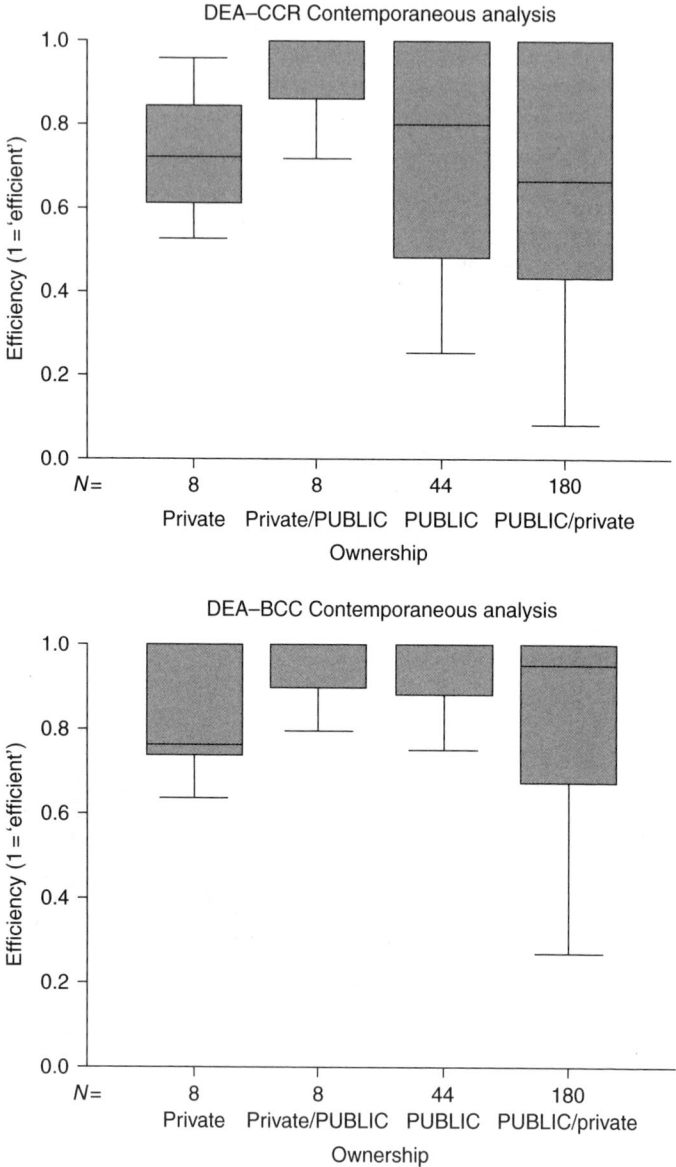

Figure 6.9   Ownership versus efficiency (1).

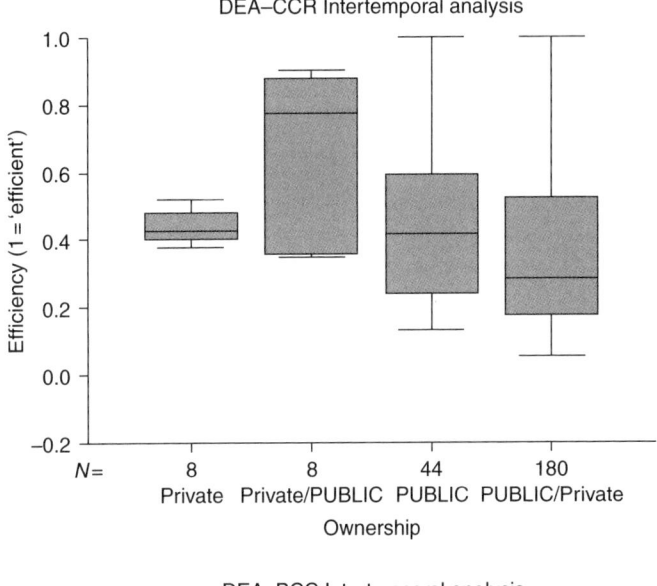

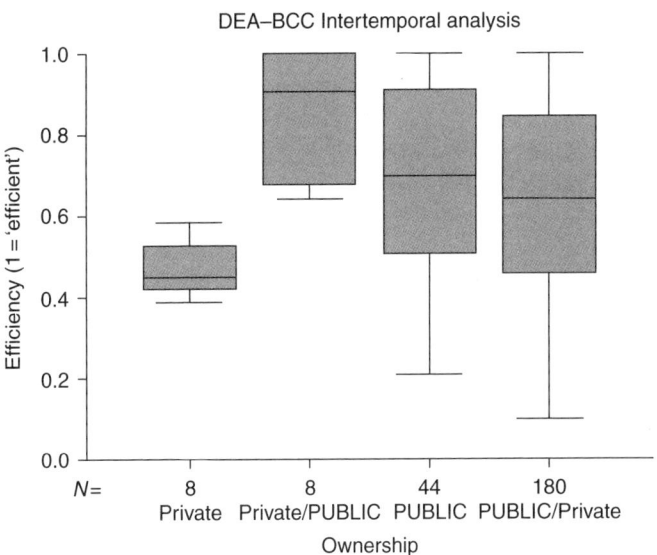

Figure 6.10   Ownership versus efficiency (2).

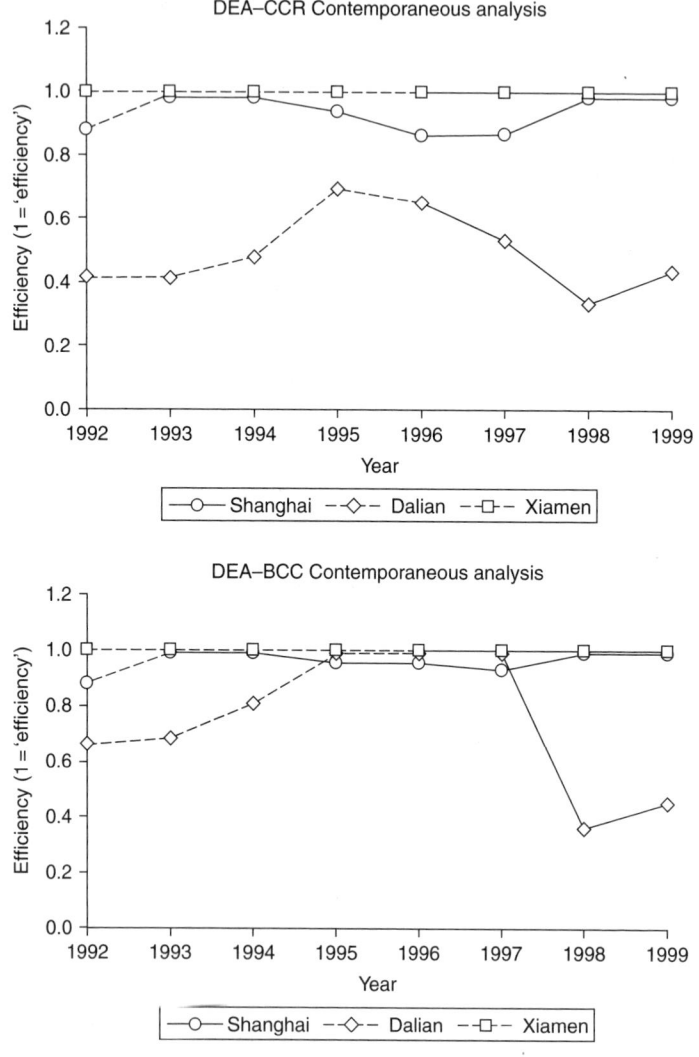

Figure 6.11 Ownership versus efficiency with reference to container ports in China* (Contemporaneous analysis).

* The dotted and solid lines stand for PUBLIC and PUBLIC/private ownership, respectively.

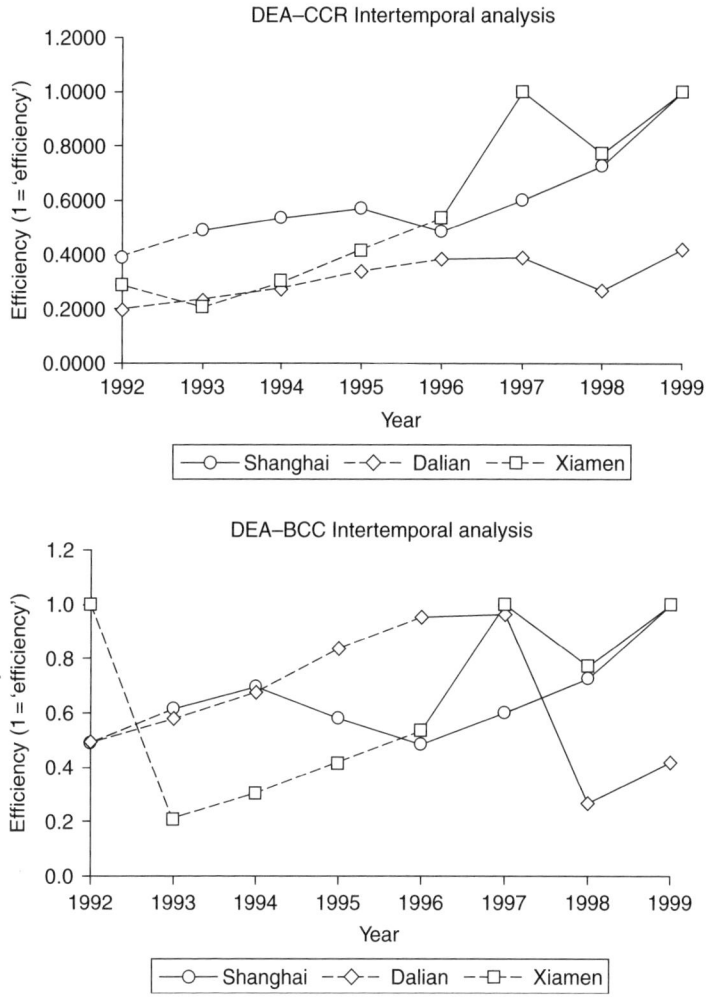

Figure 6.12 Ownership versus efficiency with reference to container ports in China* (Intertemporal analysis).

* The dotted and solid lines stand for PUBLIC and PUBLIC/private ownership, respectively.

after the change in ownership. The graphs reveal that efficiency does not exhibit any apparent immediate upward movement following the change in ownership of the ports. Perhaps, however, there has been insufficient time for any fundamental changes in the way that the ports are managed to have permeated through the two organisations so as to be reflected in efficiency improvements.

To sum up, the analysis of our sample data does not find any definitive proof to support hypothesis 5. Accordingly, it is rejected. In fact, this is a finding that is consistent with that of Cullinane and Song (2002) and Liu (1995).

Hypothesis 6: *Different firms within a port have the same or similar efficiency.*

Cross-sectional data are chosen to test hypothesis 6, because cross-sectional data contain more information on individual terminals within the sample ports than does the panel data sample. Table 6.6 is taken from Appendix 1, in which each container port comprises more than a terminal.

From a cursory glance at Table 6.6, it is clear that each container port only contains a few (rather than many) container terminals. For this reason, to calculate the mean and standard deviation of each group of terminals in the same port is not useful. As a more useful visual aid, a boxplot is employed to show the distribution of the results of the efficiency estimation of different terminals in a port and the variation between them. Figure 6.13 shows that, in most cases, the efficiency estimates of different container terminals within the same port are quite different, although the inferences that can be drawn about this aspect do, to some extent, depend on the models that have been used to derive the efficiency estimates. For instance, terminals within the ports of Antwerp and Hong Kong exhibit drastic variations in efficiency when both the DEA–CCR and DEA–BCC models are applied, but much less so when FDH models are applied. Nevertheless, it can be concluded that the sample data used within this study find no real proof to support hypothesis 6 and, accordingly, on the basis of this statistical analysis it must be rejected.

6.3   Summary

As the true focus of this book, this chapter has presented a number of empirical tests of hypotheses that have been formulated on the basis of a corpus of traditional economic theory of port production. What differentiates this work from previous studies on the subject is that both cross-sectional and panel data have been collected and analysed at the level of individual container ports or, where possible, at the level of the terminals that together comprise the port. Another distinguishing

Table 6.6 Terminal efficiency of the CCR, BCC and FDH models*

Port	Port/terminal	DEA–CCR–O	DEA–BCC–O	FDH
Hong Kong	Hong Kong	0.6922	1.0000	1.0000
	HIT	0.8267	1.0000	1.0000
	MTL	0.4942	0.6940	1.0000
	Terminal 3	1.0000	1.0000	1.0000
	Cosco-HIT	0.6200	0.7510	1.0000
Busan	Busan	0.2861	0.5435	0.5067
	Jasungdae	0.2505	0.3515	0.2581
	Shinsundae	0.3045	0.4536	0.5455
	Uam	0.3111	0.3185	0.8333
	Gamman_G	0.2686	0.6416	1.0000
	Gamman_Hanjin	0.4187	1.0000	1.0000
	Gamman_Hyundai	0.4477	1.0000	1.0000
	Gamman_K	0.3339	1.0000	1.0000
Shenzhen	Shenzhen	0.4509	0.6386	1.0000
	Yantian	0.6233	0.7578	0.8333
	Shekoui	0.4239	0.4437	0.8250
	Chiwan	0.2987	0.3067	0.8889
Hamburg	Hamburg	0.4774	0.7781	1.0000
	Burchardkai	0.7146	0.7561	1.0000
	Eurokai	0.6662	0.6688	1.0000
	TCT Tollerort	0.4789	0.6059	1.0000
	Unikai	0.1907	0.1959	0.9091
Antwerp	Antwerp	0.3310	1.0000	1.0000
	Europe Terminal	0.6424	0.6534	1.0000
	Seaport	0.1545	0.1782	0.8000
	Noord Natie	0.5481	0.7968	1.0000
	Noordzee	1.0000	1.0000	1.0000
Port Klang	Port Klang	0.2867	0.4596	0.7833
	Klang Container	1.0000	1.0000	1.0000
	Klang Port	0.2097	0.3148	0.2773
Manila	Manila	0.3686	0.5099	0.7805
	South Habour	0.3875	0.3987	0.8571
	Manila International	0.2518	0.3306	0.3200
Nagoya	Nagoya	0.6037	0.6809	0.9179
	Kinjo Pier	0.4332	1.0000	1.0000
	NCB	0.9267	1.0000	1.0000

* 1 = 'efficient'.

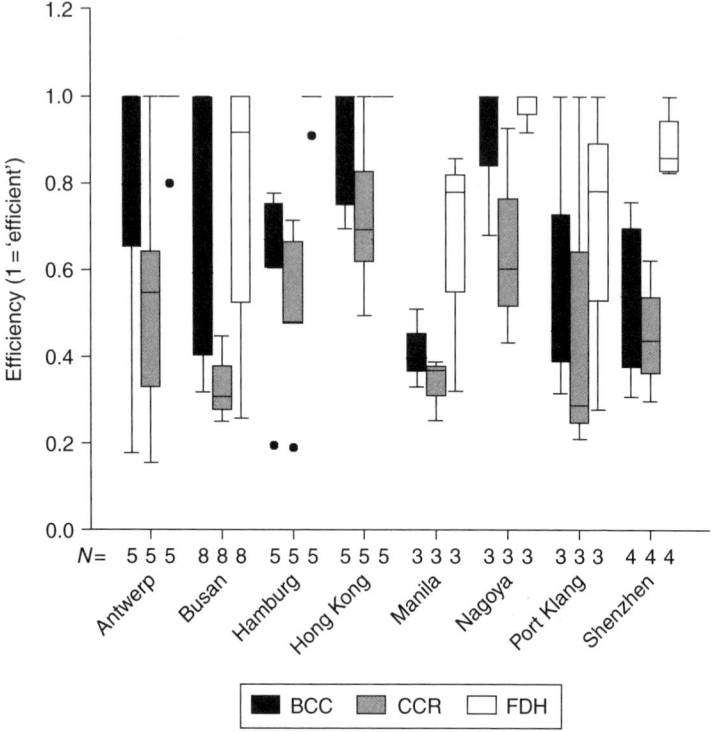

Figure 6.13   Efficiency and variation of different terminals at individual ports.

feature of the analysis contained within this book has been that it is based on a wide range of methodologies, both parametric and non-parametric, that have ensured the validity of the empirical examination that has been undertaken and the results obtained. This, however, has not been the only benefit of this analysis. It has also permitted the comparative assessment of the consistency of the results obtained from the different approaches and models and, therefore, to a large extent, has provided an empirical validation of the approaches and techniques themselves. To summarise, the main findings of this chapter specifically include:

1. As expected, the overall efficiency of the container port industry has improved with time.
2. A weak positive relationship exists between scale of production and efficiency; in other words, efficiency increases with the scale of

a container port and large container ports are more likely to be associated with high efficiency than their smaller counterparts.

3. The efficiency of larger container ports tends to be more stable than that of their smaller counterparts, although fluctuations exist in almost all container ports whether large or small in scale.

4. Inter-port competition does not necessarily improve the efficiency of container ports.

5. Port privatisation does not necessarily improve the efficiency of a port.

6. Even within the same port, the performance of different terminals may differ from each other, sometimes dramatically.

7
Conclusions and Further Research

7.1 Introduction

This chapter aims to summarise the work that has been presented within this book and, on the basis of the findings and conclusions drawn herein, to suggest further research in this field. To this end, the main contributions of this book are summarised in Section 7.2. The limitations of the research and possible further work based on the contents of the book are presented in Section 7.3.

7.2 A summary of major research contributions

The book contributes to existing studies by virtue of its thorough investigation into two important topics; the economic theories underpinning container port production and the issue of methodological choice in the estimation and measurement of technical efficiency.

7.2.1 Contribution to container port production theory

With the increasing importance of container transportation and the pivotal role in international supply chains that is played by ports, production theory as it relates to container ports merits comprehensive and rigorous theoretical and empirical investigation. Existing theory, however, fails to provide definitive insights that are directly applicable to the unique nature of the container port industry; a feature that arises from the industry's complex structure, diversity in ownership and management, multi-level objectives and operations. Against this background, this book has attempted to investigate the fundamental economic theories that are widely accepted as underpinning the container port industry and its production. The book has been structured in accordance with attempting to achieve this primary purpose.

In Chapter 2, traditional theories of economic production, in particular the theory of industrial organisation, were applied to the container port industry. A crucial differentiation was made between inter- and intra-port competition and an examination was conducted into the relationship between different sorts of competition and their impact on efficiency, as well as on the policy and managerial implications of the different configurations expounded. It was suggested that traditional economic theory might provide inappropriate explanations of the conduct and behaviour that can be observed in container port production. For example, suggestions that privatisation or competition would definitely lead to improved efficiency in container ports seemed to be somewhat contentious propositions. It was proposed that empirical evidence, derived from the application of rigorous analytical procedures, would provide the most appropriate means of addressing many of the questions that had been posed by the apparent disparity between the predictions implicit in classical economic theory and behaviour as observed within the container port industry. A number of important hypotheses were formulated for subsequent testing that related to the relationship between the technical efficiency of container ports and issues such as scale of production, asset ownership, market competition and performance. These constituted the fundamental keystones upon which the structure of the book and the investigations contained therein were based.

Chapter 3 focused on comparing alternative approaches to measuring efficiency. It emphasised the vast range of possible approaches to measuring efficiency that can be adopted, each with their own respective strengths and weaknesses that this chapter sought to identify and elaborate upon. From a methodological point of view, such an empirical comparison is of great significance, as it helps to shed light on the fundamental nature of the various available approaches, how they might be categorised into 'families' that are congruent by virtue of certain of their innate characteristics and how and in what circumstances they might actually be applied in practice. In this respect, the enumerated approaches are classified according to whether the production function frontier is non-parametric or parametric and within each of these categorisations, models were presented according to whether cross-sectional or panel data were required in order to operationalise them. It is important to reiterate that most of these approaches have never before been applied to the study of the container port industry, although they are increasingly being used in many other industries, in order to satisfy the informational requirements of a great number of policy and managerial purposes and objectives.

The justification of input and output variables in order to operationalise any or all of the different approaches that have been outlined in Chapter 2 is actually anything but the simple task it appears to be. In fact, the identification of data requirements is one of the most inherently sophisticated and difficult tasks in applying any efficiency estimation methodology. Not only must data be collected and collated in a consistent and thorough manner which facilitates the subsequent analysis, the outcomes from that analysis are fundamentally interpretable only in terms of the nature and form of the data that has constituted the input and output variables. In other words, different specifications of input and output variables not only yield different estimates of efficiency, but also lead to different implications that may be drawn from the results. In order to optimise the data collection stage of the research presented within this book, Chapter 4 involved an investigation of the internal relationship between the objectives of a firm and its input and output variables and, in consequence, allowed convincing choices of input and output variables to be made.

Chapter 5 presents the results of applying the various models to the sample data. A comprehensive comparison of the results across the various categories of model and data requirements reveals that the estimation of efficiency by non-parametric and parametric frontier approaches using cross-sectional data generates very similar results. However, in terms of achieving greater consistency in results, for this sample data at least, non-parametric frontier approaches seem to generate more convincing estimates of efficiency than their parametric counterparts. As such, non-parametric approaches, especially the various DEA techniques, constitute the fundamental basis for the further analysis in Chapter 6 of the economic theories underpinning the container port industry.

Chapter 6 provides a mirror image of the contents of Chapter 2 by presenting the main empirical results and evidence from Chapter 5 in specific relation to the original set of hypotheses that had been formulated for testing in Chapter 2 and which, in turn, had related directly to one or more of the classical elements of the economic theory that is held to underpin the workings of the container port industry. In conclusion, it was found that efficiency estimates in the container port industry does not appear to be greatly influenced by scale of production size, inter-port competition or port ownership, but does tend instead to be more firm-specific.

These findings shed light on container port production and have significant policy and managerial implications. Governments may be influenced in formulating policies with reference to the findings. For instance, the potential influence on market regulation and port privatisation are obvious given the empirical findings herein on the relationship between competition and efficiency in container ports, the relationship between port ownership and efficiency, and so on. From observing their own individual relative efficiency estimates, port or terminal operators themselves may also obtain useful insights into the state of their own productive efficiency and to learn where shortcomings exist by identifying the sources which may be contributing to their own technical inefficiency. A benchmarking analysis such as this, conducted on a purely objective and scientific basis, could then constitute the starting point for a port or terminal to further improve its efficiency, to the probable benefit of its profitability, its shareholders, society and the national economy as a whole in which it might be located.

7.2.2 Contribution to methodological choice in measuring efficiency

From a methodological point of view, very few of the existing approaches to estimating efficiency have empirical support in applications to the container port industry. Given that each approach is associated with specific strengths and weaknesses, empirical evidence is particularly important in aiding methodological choice, illustrating the process of application, highlighting operational problems and in the general evaluation of each of the various alternative approaches.

A thorough literature review revealed that most existing studies measuring the efficiency of the container port industry are limited to models using cross-sectional data. Panel data, however, have proved to be more valuable in providing deeper knowledge and insight and, in consequence, it is advocated that it should be more extensively used, wherever feasible.

Several leading approaches to measuring efficiency were compared and applied to the container port industry using both cross-sectional and panel data. As far as econometric approaches are concerned, the most widely used mathematical functional forms of a production frontier, the Cobb–Douglas production function, was chosen to represent container port production. Highly consistent efficiency estimates using cross-sectional data were found using both non-parametric and parametric approaches. On the other hand, despite the enthusiasm of econometricians, the study contained within this book has found that parametric models are still at a stage of infancy in dealing with estimating models using panel data. This is largely because of the incongruity between the complex nature of container port production and the simplicity of the models themselves. In other words, the simplified production function that is integral to any parametric approach does not yield an adequate representation of the realities of container port production. Another serious drawback with most parametric models is that they are particularly associated with strong assumptions on the distribution of efficiency and the error structure of the models employed. It is difficult, therefore, if not impossible, to determine whether these assumptions are really suitable for measuring efficiency in the container port industry. As a consequence, efficiency estimates using parametric approaches are highly questionable and, in consequence, are subject to severe criticism.

By contrast, this book finds advantages in applying non-parametric approaches to measuring container port or terminal efficiency, mainly in the form of the DEA and FDH approaches. One significant advantage of non-parametric approaches such as DEA and FDH is that they do not impose any form of functional assumptions on the efficiency estimates or error structure of this family of models. In so doing, the data are said to be able to 'speak for themselves'.

In particular, non-parametric approaches exhibit their advantages in analysing panel data. DEA-window, DEA contemporaneous and DEA Intertemporal analyses were found to be consistent with traditional production theory. Furthermore, the efficiency results revealed by these different model forms have shown that they have valuable policy implications. For instance, DEA Intertemporal analysis is capable of analysing whether a port is keeping abreast of both technological advancements and management improvements, while DEA contemporaneous analysis can theoretically 'filter' the improvements in efficiency attributed to technological advancements over time.

To sum up, such a comprehensive comparison of various approaches to estimating efficiency measures is of great significance not only in its own right, but also in preparing for the use of these and other methodologies in the further analysis of the container port industry or, for that matter, for any other industry of interest.

7.3   Recommendations for further research

Given the theoretical advantages of stochastic frontier models, as opposed to their operational difficulties, it is clear that much work needs to be done to improve and enhance the application of econometric models to analysing the efficiency of the container port industry. First, from the viewpoint of model selection, existing research is not exactly comprehensive in its treatment of the justification of model selection. The Cobb–Douglas production function was thoroughly examined in this book and applied to the container port industry. However, this book has shown that the Cobb–Douglas production function exhibits certain weaknesses in its application to the container port industry. It will be interesting to expand the boundaries of current research by applying a more extensive range of parametric models within a more comprehensive parametrically focused analysis. Second, it would be beneficial if, using stochastic frontier analysis, it would be possible to decompose efficiency estimates, especially where these change over time, into elements that are due to technological innovation (i.e. in how far a firm is keeping abreast of state-of-the-art technology) and those that are due purely to improvements in technical efficiency (i.e. how far a firm makes the most efficient use of the technology that it possesses, whether it be state-of-the-art or otherwise).

In order that the contents of this book retain their focus specifically on the issue of technical efficiency, it has ignored the financial performance of the sample container ports. Another important aspect of overall economic efficiency that the book has deliberately not addressed, however, is the influence of factor prices. Accounting for this is achieved by examining the allocative efficiency of the sample under study. An interesting potential extension to the research contained within this book, therefore, would be to examine the internal relationship between technical and allocative efficiency for the container port industry. This is especially significant in that optimum technical efficiency in the processing of inputs into outputs does not necessarily guarantee the financial success or survival of a container port.

Despite the numerous advantages it does possess, especially in comparison to some alternative approaches, the DEA methodology suffers to some extent from its inability to incorporate statistical noise. Efforts to incorporate statistical noise into DEA models are underway, but are still in a state of infancy. It will be extremely interesting to examine and apply to the container port industry any newly developed DEA approach of this kind that may come to fruition.

In the final conclusion, suffice it to say that to estimate the efficiency of a container port is the beginning and not the end of any analysis which intends to describe, define or determine a port's competitiveness and the subsequent amount of business that it might secure as a result. This study has explored the potential reasons behind efficiency (and/or inefficiency) associated merely with the sample container ports under study and has found it to be an extremely difficult, if not impossible, task to isolate or discover a single universal factor that influences the efficiency of the whole port industry. It is undoubtedly the case that each individual port has its own specific and unique context within which it operates and competes and that its level of efficiency will contribute to its operational and competitive success or otherwise. As well as applying a systematic comparative analysis as advocated and contained within this work, therefore, this implies that such an analysis needs to be supplemented by an investigation of other more singular aspects of individual ports on a case-by-case basis, not least of which will be pricing policies. It will then be extremely interesting to explore the more subtle reasons behind the degree to which each individual container port is competitive. However, before seeking to achieve those rather lofty and ambitious objectives, as this book has attempted to show, there remain numerous more fundamental questions to answer concerning the comparatively more simple influences on competitiveness, such as port pricing and efficiency.

Appendices

Appendix 1 Terminal Efficiency of the CCR, BCC and FDH Models*

Port	Port/terminal	DEA–CCR–O	DEA–BCC–O	FDH
Hong Kong	Hong Kong	0.6922	1.0000	1.0000
	HIT	0.8267	1.0000	1.0000
	MTL	0.4942	0.6940	1.0000
	Terminal 3	1.0000	1.0000	1.0000
	Cosco-HIT	0.6200	0.7510	1.0000
Singapore	Singapore	0.8999	1.0000	1.0000
Busan	Busan	0.2861	0.5435	0.5067
	Jasungdae	0.2505	0.3515	0.2581
	Shinsundae	0.3045	0.4536	0.5455
	Uam	0.3111	0.3185	0.8333
	Gamman G	0.2686	0.6416	1.0000
	Gamman Hanjin	0.4187	1.0000	1.0000
	Gamman Hyundai	0.4477	1.0000	1.0000
	Gamman K	0.3339	1.0000	1.0000
Taiwan	Kaohsiung	0.9871	1.0000	1.0000
Shanghai	Shanghai	0.7541	1.0000	1.0000
Rotterdam	Rotterdam	0.4601	0.8045	0.6452
	Home	0.7394	0.8221	1.0000
Los Angeles	Los Angeles	0.9752	1.0000	1.0000
Shenzhen	Shenzhen	0.4509	0.6386	1.0000
	Yantian	0.6233	0.7578	0.8333
	Shekoui	0.4239	0.4437	0.8250
	Chiwan	0.2987	0.3067	0.8889
Hamburg	Hamburg	0.4774	0.7781	1.0000
	Burchardkai	0.7146	0.7561	1.0000
	Eurokai	0.6662	0.6688	1.0000
	TCT Tollerort	0.4789	0.6059	1.0000
	Unikai	0.1907	0.1959	0.9091
Long Beach	Long Beach	0.9887	1.0000	1.0000
Antwerp	Antwerp	0.3310	1.0000	1.0000
	Europe Terminal	0.6424	0.6534	1.0000
	Seaport	0.1545	0.1782	0.8000
	Noord Natie	0.5481	0.7968	1.0000
	Noordzee	1.0000	1.0000	1.0000

Continued

Appendix 1 Continued

Port	Port/terminal	DEA–CCR–O	DEA–BCC–O	FDH
Port Klang	Port Klang	0.2867	0.4596	0.7833
	Klang Container	1.0000	1.0000	1.0000
	Klang Port	0.2097	0.3148	0.2773
Dubai	Dubai	0.4447	0.5526	1.0000
New York/ New Jersey	New York/ New Jersey	0.6841	1.0000	1.0000
Bremen/ Bremerhaven	Bremen/ Bremerhaven	0.6094	0.9679	0.6296
Felixstowe	Felixstowe	0.3776	0.6056	0.8226
Manila	Manila	0.3686	0.5099	0.7805
	South Habour	0.3875	0.3987	0.8571
	Manila International	0.2518	0.3306	0.3200
Tokyo	Tokyo	0.5119	0.6533	1.0000
Qingdao	Qingdao	0.4999	0.6303	0.9915
Gioia Tauro	Gioia Tauro	1.0000	1.0000	1.0000
Yokohama	Yokohama	0.3518	0.3796	1.0000
Laem Chabang	Laem Chabang	1.0000	1.0000	1.0000
Tanjung Priok	Tanjung Priok	0.5493	0.7771	1.0000
Algeciras	Algeciras	0.8992	1.0000	1.0000
Kobe	Kobe	0.2745	0.4694	0.5517
Nagoya	Nagoya	0.6037	0.6809	0.9179
	Kinjo Pier	0.4332	1.0000	1.0000
	NCB	0.9267	1.0000	1.0000
Keelung	Keelung	1.0000	1.0000	1.0000
Colombo	Colombo	1.0000	1.0000	1.0000
Dalian	Dalian	0.2295	0.3323	0.3322
Huangpu	Huangpu	0.1540	0.1674	0.7500
Nanjing	Nanjing	0.5922	1.0000	1.0000
Tianjin	Tianjin	0.6351	0.6937	1.0000
Xiamen	Xiamen	1.0000	1.0000	1.0000

* 1 = 'Efficient'.

Appendix 2   Efficiency Yielded from Alternative Parametric Approaches*

Port	Port/terminal	COLS	Half-normal	Truncated-normal	Exponential	Gamma
Hong Kong	Hong Kong	0.4809	0.7851	0.8508	0.8513	0.8606
	HIT	0.4889	0.7898	0.8545	0.8551	0.8675
	MTL	0.3969	0.7330	0.8207	0.8214	0.8338
	Terminal 3	1.0000	0.8932	0.9158	0.9160	0.9316
	Cosco-HIT	0.3763	0.7118	0.8077	0.8084	0.8472
Singapore	Singapore	0.6411	0.8455	0.8864	0.8868	0.9001
Busan	Busan	0.3090	0.6506	0.7619	0.7627	0.8034
	Jasungdae	0.2554	0.5840	0.7098	0.7107	0.7104
	Shinsundae	0.2835	0.6098	0.7309	0.7318	0.7641
	Uam	0.2762	0.6121	0.7379	0.7388	0.7323
	Gamman G	0.2544	0.5797	0.7102	0.7113	0.7100
	Gamman Hanjin	0.3966	0.7323	0.8231	0.8238	0.8323
	Gamman Hyundai	0.4185	0.7481	0.8329	0.8336	0.8681
	Gamman K	0.3259	0.6656	0.7786	0.7795	0.8173
Taiwan	Kaohsiung	0.6336	0.8396	0.8822	0.8826	0.8980
Shanghai	Shanghai	0.7718	0.8680	0.9002	0.9005	0.9064
Rotterdam	Rotterdam	0.3823	0.7265	0.8139	0.8145	0.8426
	Home	0.3961	0.7262	0.8114	0.8121	0.8471
Los Angeles	Los Angeles	0.3241	0.6635	0.7685	0.7692	0.8011
Shenzhen	Shenzhen	0.3686	0.7089	0.8042	0.8049	0.8273
	Yantian	0.4226	0.7425	0.8253	0.8260	0.8444
	Shekou	0.2967	0.6379	0.7571	0.7581	0.7779
	Chiwan	0.2912	0.6444	0.7661	0.7669	0.7720
Hamburg	Hamburg	0.3082	0.6517	0.7590	0.7598	0.7908
	Burchardkai	0.4200	0.7465	0.8257	0.8263	0.8520
	Eurokai	0.4334	0.7589	0.8357	0.8363	0.8525
	TCT Tollerort	0.2677	0.5978	0.7180	0.7189	0.7197
	Unikai	0.1620	0.4311	0.5385	0.5393	0.5475
Long Beach	Long Beach	0.4049	0.7397	0.8222	0.8228	0.8429
Antwerp	Antwerp	0.1895	0.4713	0.5664	0.5669	0.5806
	Europe Terminal	0.4324	0.7562	0.8315	0.8321	0.8667
	Seaport	0.0706	0.2292	0.2235	0.2231	0.2280
	Noord Natie	0.1700	0.4241	0.5053	0.5059	0.5110
	Noordzee	0.4465	0.7815	0.8509	0.8514	0.8675
Port Klang	Port Klang	0.2836	0.6232	0.7410	0.7418	0.7701
	Klang Container	0.4644	0.7704	0.8399	0.8405	0.8567
	Klang Port	0.2163	0.5183	0.6423	0.6433	0.6452
Dubai	Dubai	0.3755	0.7208	0.8120	0.8126	0.8467

Continued

Appendix 2   Continued

Port	Port/terminal	COLS	Half-normal	Truncated-normal	Exponential	Gamma
New York/ New Jersey	New York/ New Jersey	0.3424	0.6743	0.7708	0.7715	0.7813
Bremen/ Bremerhaven	Bremen/ Bremerhaven	0.4833	0.7884	0.8502	0.8507	0.8756
Felixstowe	Felixstowe	0.3772	0.7162	0.8098	0.8105	0.7963
Manila	Manila	0.3103	0.6609	0.7730	0.7737	0.7937
	South Habour	0.2579	0.5985	0.7270	0.7278	0.7576
	Manila International	0.2169	0.5141	0.6361	0.6370	0.6414
Tokyo	Tokyo	0.4188	0.7543	0.8333	0.8339	0.8444
Qingdao	Qingdao	0.4576	0.7715	0.8447	0.8453	0.8742
Gioia Tauro	Gioia Tauro	0.7262	0.8586	0.8923	0.8927	0.8950
Yokohama	Yokohama	0.2130	0.5195	0.6375	0.6383	0.6498
Laem Chabang	Laem Chabang	0.6494	0.8305	0.8773	0.8776	0.8845
Tanjung Priok	Tanjung Priok	0.4794	0.7805	0.8491	0.8496	0.8772
Algeciras	Algeciras	0.8288	0.8732	0.9033	0.9036	0.9122
Kobe	Kobe	0.1871	0.4812	0.5959	0.5966	0.6071
Nagoya	Nagoya	0.4007	0.7326	0.8143	0.8149	0.8267
	Kinjo Pier	0.2384	0.5634	0.6902	0.6912	0.6935
	NCB	0.5161	0.8009	0.8602	0.8607	0.8759
Keelung	Keelung	0.5317	0.8090	0.8625	0.8630	0.8821
Colombo	Colombo	0.3722	0.7259	0.8179	0.8186	0.8394
Dalian	Dalian	0.1773	0.4472	0.5531	0.5539	0.6374
Huangpu	Huangpu	0.1567	0.4134	0.5100	0.5108	0.5590
Nanjing	Nanjing	0.3043	0.6479	0.7650	0.7659	0.7987
Tianjin	Tianjin	0.2436	0.5527	0.6721	0.6731	0.6878
Xiamen	Xiamen	0.4287	0.7428	0.8252	0.8259	0.8554

* 1 = 'Efficient'.

Appendix 3 DEA–CCR Window Analysis for Container Port Efficiency*

Port	Efficiency scores								Summary measures	
	1992	1993	1994	1995	1996	1997	1998	1999	Mean	S.D.
Hong Kong	0.7211	0.7170	0.7657						0.8753	0.1152
		0.7170	0.7657	1.0000						
			0.6907	0.9702	1.0000					
				0.8935	0.9387	1.0000				
					0.9170	1.0000	1.0000			
						0.8711	0.8885	0.8997		
Singapore	0.9497	1.0000	0.6884						0.7699	0.1204
		1.0000	0.6823	0.7087						
			0.7331	0.7614	0.9129					
				0.5639	0.6760	0.7382				
					0.6732	0.7352	0.7872			
						0.7029	0.7526	0.7929		
Busan	0.7851	0.9177	1.0000						0.6696	0.2324
		0.8458	0.9810	0.8194						
			0.8554	0.6946	0.9914					
				0.4096	0.5575	0.5172				
					0.5573	0.5171	0.3642			
						0.5118	0.3526	0.3754		
Kaohsiung	0.9898	0.9258	1.0000						0.7790	0.1560
		0.8228	0.8952	1.0000						
			0.8139	0.9276	0.8893					
				0.6067	0.5820	0.6544				
					0.5732	0.6403	0.7053			
						0.5995	0.6603	0.7356		

Continued

Appendix 3 Continued

Port	Efficiency scores								Summary measures	
	1992	1993	1994	1995	1996	1997	1998	1999	Mean	S.D.
Shanghai	0.7513	0.9428	1.0000	0.9352	0.8056	0.8663	1.0000	1.0000	0.8326	0.1254
		0.8698	0.9577	0.7819	0.7084	0.8219	0.7283			
			0.8681	0.6577	0.6925	0.5986				
Rotterdam	0.3904	0.3892	0.4077	0.3544	0.4061	0.2001	0.3042	0.2745	0.2958	0.0792
		0.3436	0.3481	0.2934	0.2186	0.2001	0.3042			
			0.3012	0.1709	0.2186	0.1991				
Los Angeles	0.9802	0.9564	1.0000	1.0000	1.0000	1.0000	0.9393	0.9406	0.9758	0.0429
		0.9451	0.9862	1.0000	1.0000	1.0000	0.8298			
			0.9861	1.0000	1.0000	1.0000				
Hamburg	0.4029	0.4073	0.4762	0.3658	0.2988	0.1722	0.1942	0.1929	0.2655	0.1039
		0.3096	0.3666	0.2884	0.1607	0.1722	0.1942			
			0.2893	0.1551	0.1607	0.1722				

Long Beach									
0.6057	0.5585	0.6765	0.7478	0.6971	0.6589	1.0000		0.6807	0.1505
	0.5469	0.6733	0.7105	0.5654	0.5994	0.9295			
		0.6301	0.5847	0.5299	0.5386				
Antwerp									
0.1956	0.1994	0.2270	0.1746	0.1434	0.0795	0.0789	1.0000	0.1262	0.0515
	0.1537	0.1736	0.1382	0.0791	0.0791	0.0789			
		0.1375	0.0776	0.0788	0.0783				
Port Klang									
0.6700	0.5707	0.6978	0.5075	0.2506	0.1614	0.1684	0.0979	0.3360	0.1962
	0.4424	0.5410	0.4036	0.1351	0.1614	0.1684			
		0.4279	0.2173	0.1351	0.1614				
Dubai									
0.7821	0.7925	0.8883	0.4623	0.4217	0.2880	0.3106	0.2281	0.4531	0.2154
	0.6262	0.7019	0.3890	0.2562	0.2880	0.3081	0.3104		
		0.5534	0.2364	0.2556	0.2857				
New York/New Jersey									
0.4095	0.3984	0.4628	0.9108	0.3965	0.2543	0.1807	0.1829	0.3611	0.1990
	0.3073	0.3613	0.7603	0.2167	0.2512	0.1807			
		0.2891	0.4739	0.2141	0.2494				

Continued

154

Appendix 3 Continued

Port	Efficiency scores								Summary measures	
	1992	*1993*	*1994*	*1995*	*1996*	*1997*	*1998*	*1999*	*Mean*	*S.D.*
Bremen/ Bremerhaven	0.4658	0.4823	0.5346	0.3942	0.3214				0.2853	0.1259
		0.3499	0.3879	0.3052	0.1725	0.1692				
			0.3002	0.1638	0.1725	0.1692	0.1801			
						0.1692	0.1801	0.2167		
Felixstowe	0.6961	0.7395	0.7827	0.7202	0.6395				0.6039	0.0921
		0.6360	0.6733	0.5944	0.5382	0.5768				
			0.5769	0.5002	0.5237	0.5584	0.6313			
						0.4495	0.5186	0.5156		
Manila	0.7111	0.5436	0.6011	0.7891	0.8114				0.5517	0.1243
		0.5379	0.5733	0.6945	0.4733	0.5148				
			0.4947	0.4052	0.4731	0.5148	0.4233			
						0.5110	0.3989	0.4589		
Tokyo	0.4907	0.4236	0.4694	0.4827	0.3973				0.3219	0.1076
		0.3591	0.4031	0.4018	0.2225	0.2046				
			0.3331	0.2189	0.2225	0.2043	0.2292			
						0.2012	0.2286	0.3013		

Qingdao	0.6782	0.5415	0.8807	1.0000	1.0000	1.0000	0.7804	0.6993	0.7715	0.1809
		0.5058	0.8227	0.7407	0.7864	1.0000	0.5512			
			0.5309	0.5825	0.7864	1.0000				
Yokohama	0.5162	0.4727	0.4481	0.4502	0.3247	0.1752	0.2550	0.2318	0.3004	0.1176
		0.3836	0.3601	0.3613	0.1750	0.1752	0.2447			
			0.2885	0.1946	0.1750	0.1745				
Laem Chabang	0.1059	0.6864	0.4718	0.7085	1.0000	1.0000	1.0000	1.0000	0.6584	0.2539
		0.5510	0.4666	0.5452	0.8498	0.6414	0.8527			
			0.3591	0.4416	0.6246	0.5469				
Tanjung Priok	0.9657	0.7987	1.0000	1.0000	0.9704	1.0000	0.9479	0.8616	0.9128	0.0958
		0.6896	0.8633	1.0000	0.9273	1.0000	0.7194			
			0.8631	0.9556	0.9273	0.9394				
Kobe	0.4047	0.4213	0.3987	0.1972	0.3464	0.4824	0.4693	0.3795	0.3388	0.1046
		0.3509	0.3519	0.1769	0.2461	0.4678	0.3706			
			0.3102	0.1175	0.2457	0.3607				

Continued

Appendix 3 Continued

Port	Efficiency scores								Summary measures	
	1992	1993	1994	1995	1996	1997	1998	1999	Mean	S.D.
Nagoya	0.6979	0.7341	0.7264						0.3862	0.2019
		0.5327	0.5271	0.5611						
			0.4080	0.4343	0.4319					
				0.2331	0.2319	0.2233				
					0.2319	0.2233	0.1701			
						0.2233	0.1701	0.1911		
Keelung	1.0000	0.9566	1.0000						0.6655	0.2822
		0.8808	0.9446	1.0000						
			0.8439	0.8934	0.8681					
				0.5032	0.4890	0.3796				
					0.4873	0.3796	0.3266			
						0.3796	0.3266	0.3193		
Colombo	0.9394	1.0000	1.0000						0.9363	0.1069
		1.0000	1.0000	0.8624						
			1.0000	0.7735	1.0000					
				0.6383	0.8234	1.0000				
					0.8228	1.0000	1.0000			
						1.0000	1.0000	0.9943		
Dalian	0.3438	0.4049	0.4742						0.4510	0.0923
		0.3917	0.4587	0.5658						
			0.4235	0.5223	0.5939					
				0.4621	0.5255	0.5332				
					0.5225	0.5302	0.2916			
						0.3870	0.2687	0.4185		

City										
Huangpu	0.2552	0.2831	0.3466	0.3227	0.2845	0.2535	0.2880		0.2667	0.0361
		0.2564	0.3139	0.2765	0.2448	0.2535	0.2249		0.2895	0.1289
			0.2690	0.2379	0.2448	0.1979		0.2474		
Nanjing	0.3262	0.4854	0.5616	0.4670	0.3256	0.1747	0.1653	0.2106	0.6708	0.1387
		0.3522	0.4075	0.3615	0.1748	0.1747	0.1653			
			0.3154	0.1940	0.1748	0.1747				
Tianjin	0.3914	0.4793	0.8087	0.8967	0.7414	0.6854	0.7324	0.7763	0.8031	0.2163
		0.4604	0.8056	0.8327	0.6028	0.6728	0.6081			
			0.7481	0.6821	0.5918	0.5586				
Xiamen	0.9480	0.6877	1.0000	1.0000	1.0000	1.0000	1.0000	1.0000		
		0.4990	0.7256	0.7740	0.5368	1.0000	1.0000	0.7710		
			0.5616	0.4155	0.5368	1.0000				

* 1 = 'Efficient'.

Appendix 4 DEA–BCC Window Analysis for Container Port Efficiency*

Port	Efficiency scores								Summary measures	
	1992	1993	1994	1995	1996	1997	1998	1999	Mean	S.D.
Hong Kong	0.7226	0.7176	0.7658	1.0000	1.0000	1.0000			0.9136	0.1155
		0.7175	0.7658	0.9871	0.9707	1.0000	1.0000			
			0.9398	0.9219	0.9354	1.0000	1.0000	1.0000		
Singapore	0.9508	1.0000	1.0000	1.0000	1.0000	1.0000			0.9365	0.0770
		1.0000	0.9629	0.8344	0.9157	0.9339	1.0000			
			0.8034	0.7641	0.8552	0.8865	0.9492	1.0000		
Busan	0.7855	0.9182	1.0000	0.8738	1.0000	0.9332			0.8520	0.1579
		0.8589	0.9850	0.9308	1.0000	0.9094	0.5915			
			0.8860	0.8883	1.0000	0.7056	0.5259	0.5435		
Kaohsiung	1.0000	0.9461	1.0000	1.0000	0.9676	1.0000			0.9559	0.0527
		0.8861	0.9365	1.0000	0.9670	0.9079	1.0000			
			0.9365	1.0000	0.9455	0.8150	0.8977	1.0000		

Shanghai	0.7643	0.9591	1.0000	0.9540	0.9233	0.9281				1.0000				0.8623	0.1151					
		0.8719	0.9605	0.8522	0.8049	0.8219				0.7283										
			0.8741	0.7585	0.7224	0.5986					1.0000									
Rotterdam	0.8246	0.8205	0.8775	0.9094	0.9502	0.9463				0.8464				0.8472	0.0561					
		0.7721	0.8279	0.9105	0.8805	0.8599				0.7682										
			0.8350	0.8408	0.8011	0.7733					0.8045									
Los Angeles	1.0000	0.9900	1.0000	1.0000	1.0000	1.0000				0.9488				0.9861	0.0298					
		0.9690	0.9862	1.0000	1.0000	1.0000				0.8800										
			0.9862	1.0000	1.0000	1.0000					0.9901									
Hamburg	0.7549	0.8041	0.8130	0.8065	0.7485	0.7414				0.8136				0.7319	0.0596					
		0.7515	0.7604	0.7063	0.6757	0.7011				0.7709										
			0.6622	0.6371	0.6393	0.6570					0.7310									
Long Beach	0.8008	0.8349	1.0000	0.8859	0.8874	0.9754				1.0000				0.8536	0.1015					
		0.6479	0.8019	0.8806	0.8385	0.7942				0.9295										
			0.7971	0.8361	0.7244	0.7304					1.0000									

Continued

Appendix 4 Continued

Port	Efficiency scores								Summary measures	
	1992	1993	1994	1995	1996	1997	1998	1999	Mean	S.D.
Antwerp	0.7658	0.7827	0.9212	0.9028	0.7215	0.6588	0.7127	0.7635	0.7117	0.1043
		0.7272	0.8559	0.6458	0.6215	0.6389	0.6899			
			0.6134	0.5587	0.6140	0.6160				
Port Klang	1.0000	0.6282	0.7681	0.6946	0.4062	0.4021	0.3754	0.4596	0.5379	0.1887
		0.5720	0.6995	0.6719	0.3718	0.3723	0.3379			
			0.6749	0.5672	0.3447	0.3362				
Dubai	0.8412	0.8801	0.9865	0.4961	0.6618	0.6887	0.6846	0.5612	0.6885	0.1574
		0.8096	0.9075	0.5309	0.5966	0.6348	0.6059			
			0.8940	0.5009	0.5506	0.5618				
New York/ New Jersey	0.9058	0.8539	0.9520	1.0000	1.0000	1.0000	1.0000	1.0000	0.9224	0.0940
		0.7927	0.8821	1.0000	0.8523	1.0000	0.8993			
			0.6640	0.9485	0.8523	1.0000				

Bremen/Bremerhaven									
0.8712	0.9021	1.0000	1.0000	1.0000	0.8597	0.9148	1.0000	0.8990	0.0874
	0.8876	0.9839	0.9890	0.7790	0.8597	0.8310			
		0.9731	0.7705	0.7790	0.7810				
Felixstowe									
0.8893	0.9447	1.0000	0.7714	0.7505	0.7325	0.7407	0.5834	0.7095	0.1304
	0.6630	0.7018	0.7063	0.6817	0.6677	0.5560			
		0.6415	0.6335	0.6108	0.4963				
Manila									
0.7394	0.5476	0.6023	0.7924	0.8243	0.8783	0.5806	0.5099	0.6600	0.1316
	0.5394	0.5775	0.7056	0.8021	0.7779	0.4434			
		0.5394	0.6866	0.7917	0.5414				
Tokyo									
0.4941	0.4329	0.4816	0.5287	0.6122	0.5459	0.5468	0.6519	0.5169	0.0632
	0.3989	0.4495	0.5679	0.5755	0.5101	0.4911			
		0.4762	0.5379	0.5349	0.4688				
Qingdao									
1.0000	0.6148	1.0000	1.0000	1.0000	1.0000	0.7807	0.8442	0.8304	0.1742
	0.6148	1.0000	0.7407	0.7864	1.0000	0.6655			
		0.5309	0.5825	0.7864	1.0000				

Continued

Appendix 4 Continued

Port	Efficiency scores								Summary measures	
	1992	1993	1994	1995	1996	1997	1998	1999	Mean	S.D.
Yokohama	0.5947	0.5255	0.5496						0.5025	0.0675
		0.4928	0.5157	0.6138						
			0.5068	0.6091	0.5183					
				0.5672	0.4858	0.4824				
					0.4455	0.4426	0.4763			
						0.4015	0.4372	0.3793		
Laem Chabang	0.1542	1.0000	0.4897						0.7227	0.2703
		1.0000	0.5161	0.7837						
			0.3724	0.5655	1.0000					
				0.4613	1.0000	1.0000				
					0.6254	0.6414	1.0000			
						0.5469	0.8527	1.0000		
Tanjung Priok	1.0000	0.7987	1.0000						0.9200	0.0975
		0.6896	0.8634	1.0000						
			0.8632	1.0000	0.9704					
				0.9556	0.9273	1.0000				
					0.9273	1.0000	0.9558			
						1.0000	0.7320	0.8767		
Kobe	0.5178	0.5076	0.4947						0.4345	0.1061
		0.4806	0.4744	0.2389						
			0.4878	0.2568	0.4139					
				0.2375	0.3858	0.6265				
					0.3545	0.5175	0.5166			
						0.4173	0.4396	0.4533		

Nagoya	0.9916	1.0000	1.0000	1.0000	0.9945	0.8881	0.7360	0.8151	0.9124	0.0833
		0.9285	0.9264	1.0000	0.9024	0.8881	0.7360			
			0.9180	0.9074	0.9024	0.8881				
Keelung	1.0000	0.9566	1.0000	1.0000	0.9717	1.0000	0.8602	0.8409	0.9614	0.0538
		0.9545	0.9446	1.0000	0.9717	1.0000	0.8602			
			0.9446	1.0000	1.0000	1.0000				
Colombo	1.0000	1.0000	1.0000	1.0000	1.0000	1.0000	1.0000	0.9943	0.9506	0.1007
		1.0000	1.0000	0.7965	0.8317	1.0000	1.0000			
			1.0000	0.6561	0.8315	1.0000				
Dalian	0.5487	0.6464	0.7570	0.9220	1.0000	1.0000	0.2926	0.4198	0.7595	0.2465
		0.6383	0.7476	0.8794	0.9854	1.0000	0.2695			
			0.7130	0.8666	0.9854	1.0000				
Huangpu	0.7363	0.8169	1.0000	1.0000	1.0000	1.0000	1.0000	1.0000	0.9277	0.0904
		0.7945	0.9726	1.0000	0.9659	0.8800	0.9091			
			0.9726	1.0000	0.8500	0.8000				

Continued

Appendix 4 Continued

Port	Efficiency scores								Summary measures	
	1992	1993	1994	1995	1996	1997	1998	1999	Mean	S.D.
Nanjing	0.3262	0.4854	0.5616	0.4670	0.3256	0.1747	0.1653		0.2895	0.1289
		0.3522	0.4075	0.3615	0.1748	0.1747	0.1653			
			0.3154	0.1940	0.1748	0.1747		0.2106		
Tianjin	0.3969	0.4860	0.8534	1.0000	0.7694	0.7507	0.7330		0.7295	0.1882
		0.4739	0.8984	1.0000	0.6603	0.6734	0.6087			
			0.8984	1.0000	0.5923	0.5591		0.7770		
Xiamen	1.0000	0.6877	1.0000	1.0000	1.0000	1.0000	1.0000		0.8060	0.2187
		0.4990	0.7256	0.7740	0.5368	1.0000	0.7710			
			0.5616	0.4155	0.5368	1.0000		1.0000		

* 1 = 'Efficient'.

Appendix 5   DEA–CCR Contemporaneous Analysis of Container Port Efficiency*

	1992	1993	1994	1995	1996	1997	1998	1999	Average
Hong Kong	0.7877	0.7170	1.0000	1.0000	1.0000	1.0000	1.0000	0.9308	0.9294
Singapore	1.0000	1.0000	1.0000	0.8392	0.9237	0.7382	1.0000	1.0000	0.9376
Busan	0.8808	1.0000	1.0000	0.8954	0.9914	0.5172	0.6362	0.4903	0.8014
Kaohsiung	1.0000	1.0000	1.0000	1.0000	0.8919	0.6544	1.0000	1.0000	0.9433
Shanghai	0.8771	1.0000	1.0000	0.9352	0.8607	0.8663	1.0000	1.0000	0.9424
Rotterdam	0.4141	0.4198	0.4125	0.3544	0.4061	0.2001	0.7013	0.5633	0.4340
Los Angeles	1.0000	1.0000	1.0000	1.0000	1.0000	1.0000	0.9609	0.9752	0.9920
Hamburg	0.4332	0.5131	0.4762	0.3703	0.2988	0.1722	0.7229	0.6444	0.4539
Long Beach	0.6800	0.5811	0.6907	0.8786	0.7052	0.6589	1.0000	1.0000	0.7743
Antwerp	0.2867	0.2546	0.2270	0.1746	0.1434	0.0798	0.3564	0.4315	0.2443
Port Klang	0.8371	0.6989	0.6978	0.5167	0.2506	0.1614	0.3966	0.4187	0.4972
Dubai	0.9027	0.9674	0.8883	0.4882	0.4223	0.2880	0.7288	0.5717	0.6572
New York/ New Jersey	0.5253	0.4634	0.4628	0.9108	0.3965	0.2543	0.7236	0.9224	0.5824
Bremen/ Bremerhaven	0.4913	0.6735	0.5346	0.3942	0.3214	0.1692	0.9453	1.0000	0.5662
Felixstowe	0.9612	0.9086	0.7827	0.7311	0.6466	0.5768	0.7102	0.5269	0.7305
Manila	0.7488	0.5590	0.6468	0.8871	0.8114	0.5148	0.5842	0.4790	0.6539
Tokyo	0.5397	0.4833	0.5044	0.5138	0.3973	0.2046	0.5861	0.6710	0.4875
Qingdao	0.7897	0.6287	0.8807	1.0000	1.0000	1.0000	0.8241	0.7017	0.8531
Yokohama	0.5806	0.5601	0.4481	0.4615	0.3247	0.1752	0.4770	0.3732	0.4251
Laem Chabang	0.1461	0.8594	0.4827	0.8477	1.0000	1.0000	1.0000	1.0000	0.7920
Tanjung Priok	1.0000	1.0000	1.0000	1.0000	1.0000	1.0000	1.0000	0.8735	0.9842
Kobe	0.4671	0.4865	0.4304	0.2112	0.3634	0.4824	0.4930	0.3795	0.4142
Nagoya	0.7362	1.0000	0.7264	0.5611	0.4319	0.2233	0.8927	1.0000	0.6965
Keelung	1.0000	1.0000	1.0000	1.0000	0.8781	0.3796	1.0000	1.0000	0.9072
Colombo	1.0000	1.0000	1.0000	0.9057	1.0000	1.0000	1.0000	1.0000	0.9882
Dalian	0.4189	0.4121	0.4804	0.6965	0.6519	0.5332	0.3362	0.4367	0.4957
Huangpu	0.3530	0.3352	0.3466	0.3619	0.2991	0.2535	0.3116	0.2520	0.3141
Nanjing	0.3440	0.6612	0.5616	0.4670	0.3256	0.1747	0.8676	1.0000	0.5502
Tianjin	0.4789	0.4864	0.8245	1.0000	0.7646	0.6854	0.8125	0.7763	0.7286
Xiamen	1.0000	1.0000	1.0000	1.0000	1.0000	1.0000	1.0000	1.0000	1.0000
Average	0.6893	0.7223	0.7168	0.7134	0.6502	0.5321	0.7689	0.7473	0.6926
Number of efficient ports	7	10	10	8	7	7	10	11	

* 1 = 'Efficient'.

Appendix 6   DEA–BCC Contemporaneous Analysis of Container Port Efficiency*

	1992	1993	1994	1995	1996	1997	1998	1999	Average
Hong Kong	0.7953	0.7189	1.0000	1.0000	1.0000	1.0000	1.0000	1.0000	0.9393
Singapore	1.0000	1.0000	1.0000	1.0000	1.0000	1.0000	1.0000	1.0000	1.0000
Busan	0.9108	1.0000	1.0000	1.0000	1.0000	1.0000	0.6384	0.5435	0.8866
Kaohsiung	1.0000	1.0000	1.0000	1.0000	1.0000	1.0000	1.0000	1.0000	1.0000
Shanghai	0.8771	1.0000	1.0000	0.9540	0.9533	0.9281	1.0000	1.0000	0.9641
Rotterdam	0.9501	0.8636	0.8997	0.9117	0.9911	0.9463	0.8464	0.8045	0.9017
Los Angeles	1.0000	1.0000	1.0000	1.0000	1.0000	1.0000	0.9644	1.0000	0.9956
Hamburg	0.7688	0.8671	0.8148	0.8065	0.7704	0.7414	0.9398	0.7876	0.8120
Long Beach	0.8710	0.9599	1.0000	0.8871	0.9529	0.9997	1.0000	1.0000	0.9588
Antwerp	0.7954	0.8555	0.9219	0.9028	0.7255	0.6588	0.9079	0.8839	0.8314
Port Klang	1.0000	0.7046	0.7887	0.6946	0.4111	0.4029	0.4107	0.4596	0.6090
Dubai	0.9307	0.9717	1.0000	0.5492	0.6753	0.6887	0.7337	0.5751	0.7655
New York/ New Jersey	0.9482	0.9353	0.9522	1.0000	1.0000	1.0000	1.0000	1.0000	0.9795
Bremen/ Bremerhaven	1.0000	1.0000	1.0000	1.0000	1.0000	0.8597	1.0000	1.0000	0.9825
Felixstowe	1.0000	1.0000	1.0000	0.7796	0.7505	0.7356	0.7418	0.6376	0.8306
Manila	0.7754	0.5689	0.6469	0.8894	0.8243	0.9013	0.5846	0.5099	0.7126
Tokyo	0.5805	0.4856	0.5283	0.5850	0.6302	0.5501	0.6022	0.6717	0.5792
Qingdao	1.0000	1.0000	1.0000	1.0000	1.0000	1.0000	0.8890	0.8674	0.9695
Yokohama	0.6570	0.5914	0.5505	0.6138	0.5417	0.4824	0.4929	0.3847	0.5393
Laem Chabang	1.0000	1.0000	0.5639	1.0000	1.0000	1.0000	1.0000	1.0000	0.9455
Tanjung Priok	1.0000	1.0000	1.0000	1.0000	1.0000	1.0000	1.0000	0.8860	0.9857
Kobe	0.5969	0.5314	0.5368	0.2681	0.4251	0.6265	0.5253	0.5040	0.5018
Nagoya	1.0000	1.0000	1.0000	1.0000	1.0000	0.8881	0.8932	1.0000	0.9727
Keelung	1.0000	1.0000	1.0000	1.0000	1.0000	1.0000	1.0000	1.0000	1.0000
Colombo	1.0000	1.0000	1.0000	1.0000	1.0000	1.0000	1.0000	1.0000	1.0000
Dalian	0.6631	0.6829	0.8099	1.0000	1.0000	1.0000	0.3635	0.4520	0.7464
Huangpu	1.0000	1.0000	1.0000	1.0000	1.0000	1.0000	1.0000	1.0000	1.0000
Nanjing	0.4010	0.7059	0.5616	0.4670	0.3256	0.1747	1.0000	1.0000	0.5795
Tianjin	0.4943	0.5171	1.0000	1.0000	0.8151	0.7507	0.8152	0.7770	0.7712
Xiamen	1.0000	1.0000	1.0000	1.0000	1.0000	1.0000	1.0000	1.0000	1.0000
Average	0.8672	0.8653	0.8858	0.8770	0.8597	0.8445	0.8450	0.8248	0.8587
Number of efficient ports	14	15	18	17	16	14	14	15	

* 1 = 'Efficient'.

Appendix 7 DEA–CCR Intertemporal Analysis of Container Port Efficiency*

	1992	1993	1994	1995	1996	1997	1998	1999	*Average*
Hong Kong	0.3500	0.3480	0.3716	0.7575	0.7924	0.8711	0.8885	0.8997	**0.6599**
Singapore	0.6921	0.7990	0.5171	0.5370	0.6436	0.7029	0.7526	0.7929	**0.6796**
Busan	0.3329	0.3892	0.4659	0.4062	0.5532	0.5118	0.3526	0.3754	**0.4234**
Kaohsiung	0.4244	0.4339	0.5020	0.5838	0.5612	0.5995	0.6603	0.7356	**0.5626**
Shanghai	0.3892	0.4883	0.5368	0.5675	0.4839	0.5986	0.7283	1.0000	**0.5991**
Rotterdam	0.1736	0.1691	0.1688	0.1679	0.2186	0.1991	0.3042	0.2745	**0.2095**
Los Angeles	0.7416	0.7574	0.9401	0.9687	1.0000	1.0000	0.8298	0.9406	**0.8973**
Hamburg	0.1295	0.1308	0.1556	0.1551	0.1607	0.1722	0.1942	0.1929	**0.1614**
Long Beach	0.3833	0.3509	0.4220	0.4796	0.4823	0.5386	0.9295	1.0000	**0.5733**
Antwerp	0.0670	0.0684	0.0772	0.0775	0.0784	0.0783	0.0789	0.0979	**0.0780**
Port Klang	0.2020	0.1883	0.2302	0.2173	0.1351	0.1614	0.1684	0.2281	**0.1913**
Dubai	0.2731	0.2834	0.3176	0.2299	0.2492	0.2857	0.3081	0.3104	**0.2822**
New York/ New Jersey	0.1460	0.1410	0.1650	0.4736	0.2125	0.2494	0.1807	0.1829	**0.2189**
Bremen/ Bremerhaven	0.1404	0.1454	0.1612	0.1638	0.1725	0.1692	0.1801	0.2167	**0.1687**
Felixstowe	0.3796	0.4032	0.4268	0.3999	0.4303	0.4495	0.5186	0.5156	**0.4405**
Manila	0.3452	0.2780	0.2998	0.3994	0.4666	0.5110	0.3989	0.4589	**0.3947**
Tokyo	0.1855	0.1587	0.1852	0.2184	0.2218	0.2012	0.2286	0.3013	**0.2126**
Qingdao	0.4057	0.2567	0.4175	0.5825	0.7864	1.0000	0.5512	0.6993	**0.5874**
Yokohama	0.1745	0.1662	0.1554	0.1946	0.1750	0.1745	0.2447	0.2318	**0.1896**
Laem Chabang	0.0556	0.3607	0.2406	0.3653	0.5371	0.5469	0.8527	1.0000	**0.4949**
Tanjung Priok	0.5782	0.6128	0.7672	0.8977	0.8711	0.9394	0.7194	0.8616	**0.7809**
Kobe	0.1514	0.1566	0.1813	0.1042	0.2098	0.3607	0.3706	0.3795	**0.2393**
Nagoya	0.2104	0.2213	0.2190	0.2331	0.2319	0.2233	0.1701	0.1911	**0.2125**
Keelung	0.4405	0.4214	0.4652	0.4925	0.4786	0.3796	0.3266	0.3193	**0.4155**
Colombo	0.7081	0.7575	0.7485	0.6257	0.8143	1.0000	1.0000	0.9943	**0.8311**
Dalian	0.1971	0.2322	0.2719	0.3354	0.3814	0.3870	0.2687	0.4185	**0.3115**
Huangpu	0.1334	0.1480	0.1812	0.1863	0.1912	0.1979	0.2249	0.2474	**0.1888**
Nanjing	0.0983	0.1463	0.1693	0.1940	0.1748	0.1747	0.1653	0.2106	**0.1667**
Tianjin	0.2155	0.2639	0.4678	0.5207	0.4914	0.5586	0.6081	0.7763	**0.4878**
Xiamen	0.2858	0.2073	0.3015	0.4155	0.5368	1.0000	0.7710	1.0000	**0.5647**
Average	**0.3003**	**0.3161**	**0.3510**	**0.3984**	**0.4247**	**0.4747**	**0.4659**	**0.5284**	**0.4074**
Number of efficient ports	0	0	0	0	1	4	1	4	

* 1 = 'Efficient'.

Appendix 8   DEA–BCC Intertemporal Analysis of Container Port Efficiency*

	1992	1993	1994	1995	1996	1997	1998	1999	Average
Hong Kong	0.6393	0.6463	0.7107	0.8845	0.9285	1.0000	1.0000	1.0000	**0.8512**
Singapore	0.9152	1.0000	0.6522	0.6774	0.8118	0.8865	0.9492	1.0000	**0.8615**
Busan	0.5348	0.6251	0.6792	0.7202	0.7876	0.7056	0.5000	0.5398	**0.6365**
Kaohsiung	0.7514	0.6749	0.7608	0.8636	0.8309	0.8150	0.8977	1.0000	**0.8243**
Shanghai	0.4908	0.6159	0.6980	0.5810	0.4841	0.5986	0.7283	1.0000	**0.6496**
Rotterdam	0.5856	0.5856	0.6310	0.6853	0.7206	0.7733	0.7682	0.8045	**0.6943**
Los Angeles	0.7775	0.7860	0.9536	0.9831	1.0000	1.0000	0.8800	0.9901	**0.9213**
Hamburg	0.4872	0.5191	0.5303	0.5655	0.5994	0.6570	0.7709	0.7310	**0.6075**
Long Beach	0.4394	0.4515	0.5529	0.6174	0.6400	0.7304	0.9295	1.0000	**0.6701**
Antwerp	0.4320	0.4376	0.5150	0.5421	0.6049	0.6160	0.6899	0.7635	**0.5751**
Port Klang	0.6022	0.4205	0.5141	0.4999	0.2991	0.3330	0.3322	0.4463	**0.4309**
Dubai	0.5949	0.6442	0.7221	0.4040	0.4843	0.5604	0.6043	0.5607	**0.5719**
New York/ New Jersey	0.5295	0.5011	0.5821	0.9479	0.8523	1.0000	0.8993	1.0000	**0.7890**
Bremen/ Bremerhaven	0.6287	0.6510	0.7217	0.7335	0.7599	0.7810	0.8310	1.0000	**0.7634**
Felixstowe	0.3893	0.4136	0.4378	0.4287	0.4613	0.4963	0.5560	0.5834	**0.4708**
Manila	0.4645	0.3583	0.4028	0.5219	0.6149	0.5414	0.4434	0.5099	**0.4821**
Tokyo	0.3597	0.3235	0.3724	0.4399	0.4804	0.4611	0.4881	0.6476	**0.4466**
Qingdao	1.0000	0.2567	0.4175	0.5825	0.7864	1.0000	0.6555	0.8315	**0.6913**
Yokohama	0.4539	0.3837	0.3928	0.4719	0.4039	0.4015	0.4372	0.3784	**0.4154**
Laem Chabang	0.1542	1.0000	0.2465	0.3743	0.5398	0.5469	0.8527	1.0000	**0.5893**
Tanjung Priok	0.8806	0.6574	0.8231	0.9556	0.9273	1.0000	0.7320	0.8767	**0.8566**
Kobe	0.3690	0.3732	0.3727	0.1980	0.3124	0.4173	0.4396	0.4533	**0.3669**
Nagoya	0.7564	0.7956	0.8106	0.9074	0.9024	0.8881	0.7360	0.8151	**0.8265**
Keelung	0.9435	0.9026	0.9446	1.0000	0.9717	1.0000	0.8602	0.8409	**0.9329**
Colombo	1.0000	1.0000	1.0000	0.6518	0.8228	1.0000	1.0000	0.9943	**0.9336**
Dalian	0.4916	0.5790	0.6781	0.8364	0.9511	0.9651	0.2695	0.4198	**0.6488**
Huangpu	0.7161	0.7945	0.9726	1.0000	0.7727	0.8000	0.9091	1.0000	**0.8706**
Nanjing	0.0983	0.1463	0.1693	0.1940	0.1748	0.1747	0.1653	0.2106	**0.1667**
Tianjin	0.2166	0.2653	0.4682	0.5211	0.4918	0.5591	0.6087	0.7770	**0.4885**
Xiamen	1.0000	0.2073	0.3015	0.4155	0.5368	1.0000	0.7710	1.0000	**0.6540**
Average	0.5901	0.5672	0.6011	0.6401	0.6651	0.7236	0.6902	0.7725	**0.6562**
Number of efficient ports	3	3	1	2	1	8	2	10	

* 1 = 'Efficient'.

Appendix 9 Technical-invariant Efficiency Estimated by Alternative Models*

Port	Fixed effects	Random effects		
	Within	*GLS*	*MLE-half-normal*	*MLE-exponential*
Hong Kong	0.5618	0.7544	0.5314	0.8336
Singapore	0.6027	1.0000	0.5493	0.8134
Busan	0.5477	0.8340	0.5333	0.8321
Kaohusiung	0.4672	0.8701	0.5379	0.8214
Shanghai	0.5306	0.5220	0.4953	0.8660
Rotterdam	0.1594	0.4305	0.3748	0.9199
Los Angeles	0.1997	0.3288	0.4146	0.9043
Hamburg	0.1766	0.4833	0.3905	0.9153
Long Beach	0.2160	0.3148	0.4032	0.9099
Antwerp	0.0647	0.2311	0.2471	0.9460
Port Klang	0.2747	0.5088	0.4094	0.9107
Dubai	0.3530	0.6090	0.4658	0.8847
New York/New Jersey	0.1642	0.4934	0.4242	0.9002
Bremen/Bremerhaven	0.0787	0.3220	0.3185	0.9308
Flexistowe	0.4154	0.4357	0.4646	0.8859
Manila	0.1734	0.2952	0.3606	0.9237
Tokyo	0.2507	0.4569	0.3949	0.9154
Qingdao	0.6201	0.5508	0.5065	0.8583
Yokohama	0.2014	0.4162	0.3759	0.9204
Laem	0.4115	0.4880	0.5192	0.8445
Tanjung Priok	0.6046	0.5340	0.4928	0.8693
Kobe	0.1218	0.2190	0.2767	0.9421
Nagoya	0.2530	0.6865	0.4648	0.8796
Keelung	0.2667	0.5465	0.4337	0.8968
Colombo	0.3798	0.3681	0.4132	0.9089
Dalian	0.2105	0.1973	0.2929	0.9401
Huangpu	0.2040	0.1641	0.2558	0.9469
Nanjing	0.2172	0.3198	0.3278	0.9330
Tianjin	0.2681	0.2803	0.3651	0.9239
Xiamen	1.0000	0.8844	0.5554	0.8042
Average	**0.3332**	**0.4848**	**0.4198**	**0.8927**

* 1 = 'Efficient'.

Appendix 10   Time-varying Efficiency by MLE*

	1992	*1993*	*1994*	*1995*	*1996*	*1997*	*1998*	*1999*	*Average*
Hong Kong	0.6351	0.6336	0.6652	0.6674	0.6839	0.7024	0.7091	0.7326	**0.6787**
Singapore	0.7027	0.7303	0.5586	0.5741	0.6472	0.6809	0.7057	0.7580	**0.6697**
Busan	0.6246	0.6851	0.7312	0.6793	0.7760	0.7254	0.5388	0.5558	**0.6645**
Kaohsiung	0.5495	0.5701	0.6629	0.7123	0.6993	0.6872	0.7212	0.7554	**0.6697**
Shanghai	0.4850	0.5756	0.6047	0.6130	0.6010	0.6869	0.7544	0.8370	**0.6447**
Rotterdam	0.5042	0.5099	0.5456	0.6114	0.7378	0.7001	0.7940	0.7748	**0.6472**
Los Angeles	0.6322	0.6388	0.6830	0.7057	0.7244	0.6340	0.6803	0.7248	**0.6779**
Hamburg	0.5984	0.6055	0.6570	0.6606	0.6757	0.7024	0.7498	0.7401	**0.6737**
Long Beach	0.5201	0.5282	0.5980	0.6546	0.6762	0.7012	0.7731	0.7922	**0.6555**
Antwerp	0.5739	0.5798	0.6418	0.6595	0.7047	0.6898	0.7217	0.7793	**0.6688**
Port Klang	0.7804	0.6804	0.7489	0.7259	0.4960	0.5672	0.5720	0.6958	**0.6583**
Dubai	0.7034	0.7128	0.7499	0.5678	0.5988	0.6614	0.6898	0.7006	**0.6731**
New York/ New Jersey	0.5050	0.4947	0.5450	0.6512	0.7504	0.7942	0.6865	0.7593	**0.6483**
Bremen/ Bremerhaven	0.6055	0.6197	0.6605	0.6668	0.6830	0.6869	0.7091	0.7662	**0.6747**
Felixstowe	0.5738	0.5986	0.6216	0.6608	0.6866	0.7099	0.7519	0.7553	**0.6698**
Manila	0.6738	0.5868	0.6182	0.7302	0.7795	0.7306	0.5713	0.6433	**0.6667**
Tokyo	0.6336	0.5725	0.6171	0.6832	0.6959	0.6477	0.7085	0.7917	**0.6688**
Qingdao	0.4371	0.3767	0.5537	0.6669	0.7815	0.8349	0.6325	0.7210	**0.6255**
Yokohama	0.7023	0.6710	0.6549	0.7316	0.6908	0.6758	0.6772	0.6329	**0.6796**
Laem Chabang	0.0526	0.2432	0.3702	0.5174	0.7103	0.7739	0.8632	0.8831	**0.5518**
Tanjung Priok	0.6107	0.5364	0.6278	0.6869	0.6758	0.7034	0.7447	0.7625	**0.6685**
Kobe	0.7219	0.7334	0.7146	0.4671	0.7045	0.6520	0.6625	0.6942	**0.6688**
Nagoya	0.6594	0.6792	0.6809	0.7137	0.7123	0.7046	0.6218	0.6642	**0.6795**
Keelung	0.6830	0.6663	0.6493	0.6838	0.6731	0.7429	0.6796	0.6588	**0.6796**
Colombo	0.6648	0.7396	0.7351	0.5252	0.6380	0.7173	0.6783	0.6746	**0.6716**
Dalian	0.5266	0.5932	0.6565	0.7318	0.7706	0.7746	0.5099	0.6865	**0.6562**
Huangpu	0.5673	0.6098	0.6886	0.6986	0.5862	0.6004	0.7612	0.7872	**0.6624**
Nanjing	0.4688	0.6276	0.6841	0.7316	0.6957	0.6956	0.6751	0.7570	**0.6669**
Tianjin	0.5932	0.6736	0.5697	0.6134	0.6501	0.6983	0.7275	0.7972	**0.6654**
Xiamen	0.4585	0.3993	0.5372	0.6663	0.7543	0.8774	0.5248	0.6304	**0.6060**
Average	**0.5816**	**0.5957**	**0.6344**	**0.6553**	**0.6886**	**0.7053**	**0.6865**	**0.7304**	**0.6597**
Number of efficient ports	0	0	0	0	0	0	0	0	0

* 1 = 'Efficient'.

References

Afriat, S. (1972) Efficiency Estimation of Production Functions, *International Economic Review*, 13(3), 568–598.

Ahmad, M. and Bravo-Ureta, B. E. (1996) Technical Efficiency Measures for Dairy Farms Using Panel Data: A Comparison of Alternative Model Specifications, *Journal of Productivity Analysis*, 7, 399–415.

Aigner, D. and Chu, S. (1968) On Estimating the Industry Production Function, *American Economic Review*, 58(4), 826–839.

Aigner, D., Lovell, C. and Schmidt, P. (1977) Formulation and Estimation of Stochastic Frontier Production Function Models, *Journal of Econometrics*, 6(1), 21–37.

Alam, M. S. (1984) Hirschman's Taxonomy of Industries: Some Hypotheses and Evidence, *Economic Development and Cultural Change*, 32(2), 367–372.

Alberghini, G. (2002) *The Potential Benefits of a European Intermodal Network for the Transport of Maritime Containers*, Maritime & Port Symposium, 18–20 April, Antwerp.

Alderton, P. (1999) *Port Management and Operation*, London: LLP Reference Publication.

Ali, A. I. and Seiford, L. M. (1993) The Mathematical Programming Approach to Efficiency Analysis, in Fried, H. O., Lovell, C. A. K. and Schmidt, S. S. (eds), *The Measurement of Productive Efficiency: Techniques and Applications*, Oxford: Oxford University Press, 160–194.

Allen, R., Athanassopoulos, A., Dyson, R. G. and Thanassoulis, E. (1997) Weights Restrictions and Value Judgements in Data Envelopment Analysis: Evolution, Development and Future Directions, *Annals of Operational Research*, 73, 13–34.

Amato, L. (1995) The Choice of Structure Measure in Industrial Economics, *Quarterly Journal of Business and Economics*, 34(2), 39–52.

Amin, A. (1994) Post-Fordism: Models, Fantasies and Phantoms of Transition, in Amin, A. (ed.), *Post-Fordism: A Reader*, Oxford: Blackwell.

Ashar, A. (1995) Factor Analysis and Benchmarking Ports' Performance, *Maritime Policy and Management*, 22(4), 389–390.

Ashar, A. (1997) Counting the Moves, *Port Development International*, November, 25–29.

Avriel, M. and Penn, M. (1993) Exact and Approximate Solutions of the Container Ship Stowage Problem, *Computers and Industrial Engineering*, 25, 271–274.

Avriel, M., Penn, M. and Shpirer, N. (2000) Container Ship Stowage Problem: Complexity and Connection to the Colouring of Circle Graphs, *Discrete Applied Mathematics*, 103, 271–279.

Baird, A. (1995) UK Port Privatisation: in Context, in: *Proceedings of UK Port Privatisation Conference, Scottish Transport Studies Group*, 21 September, Edinburgh.

Baird, A. (1997) Port Privatisation: An Analytical Framework, in: *Proceedings of International Association of Maritime Economist Conference*, 22–24 September, City University, London.

Banker, R. D. (1996) Hypothesis Tests Using Data Envelopment Analysis, *Journal of Productivity Analysis*, 7, 139–159.

Banker, R. D., Charnes, A. and Cooper, W. W. (1984) Some Models for Estimating Technical and Scale Inefficiencies in Data Envelopment Analysis, *Management Science*, 30(2), 1078–1092.

Banker, R. D., Conrad, R. F. and Strauss, R. P. (1986) A Comparative Application of Data Envelopment Analysis and Translog Methods: An Illustrative Study of Hospital Production, *Management Science*, 32(1), 30–44.

Barros, C. P. and Athanassiou, M. (2004) Efficiency in European Seaports with DEA: Evidence from Greece and Portugal, *Maritime Economics & Logistics*, 6(2), 122–140.

Bauer, P. (1990) Recent Development in the Econometric Estimation of Frontiers, *Journal of Econometrics*, 46, 39–56.

Bauer, P. W. and Hancock, D. (1993) The Efficiency of the Federal Reserve in Providing Check Processing Services, *Journal of Banking & Finance*, 17(2–3), 287–311.

Baumol, W., Panzar, J. and Willig, R. (1982) *Contestable Markets and the Theory of Industry Structure*, New York: Harcourt, Bruce and Jovandich, Inc.

Beattie, B. R. and Taylor, C. R. (1985) *The Economics of Production*, New York: John Wiley and Sons.

Bendall, H. and Stent, A. (1987) On Measuring Cargo Handling Productivity, *Maritime Policy and Management*, 14(4), 337–343.

Bernard, J. (1991) *European Deep-Sea Container Terminals: Locational and Operational Perspectives*, PhD Thesis, University of Liverpool.

Bischoff, E. E. and Marriott, M. D. (1990) A comparative Evaluation of Heuristics for Container Loading, *European Journal of Operations Research*, 44, 267–276.

Bowlin, W. F., Charnes, A., Cooper, W. W. and Sherman, H. D. (1985) Data Envelopment Analysis and Regression Approaches to Efficiency Estimation and Evaluation, *Annals of Operations Research*, 2(1), 113–138.

Braeutigam, R., Daughety, A. and Turnquist, M. (1984) A Firm Specific Analysis of Economies of Density in the US Railroad Industry, *Journal of Industrial Economics*, 33, 3–20.

Brandenburger, A. M. and Nalebuff, B. J. (1996) *Co-opetition*, New York: Currency.

Breusch, T., Mizon, G. and Schmidt, P. (1989) Efficient Estimation Using Panel Data, *Econometrica*, 57, 695–700.

Brooks, M. (2000) *Sea Change in Liner Shipping*, Oxford: Elsevier.

Brooks, M. (2002) Issues in Measuring the Performance of Devolution Programs, *Comparative Corporate Governance in The Port Sector, Second Meeting of the Port Performance Research Network*, 12 November, Panama.

Case, K. E. and Fair, R. C. (1999) *Principles of Economics*, 5th edn, Upper Saddle River, NJ: Prentice-Hall.

Chang, S. (1978) Production Function, Productivities and Capacity Utilisation of the Port of Mobile, *Maritime Policy and Management*, 5, 297–305.

Charnes, A., Cooper, W. W. and Rhodes, E. (1978) Measuring the Efficiency of Decision Making Units, *European Journal of Operational Research*, 2, 429–444.

Charnes, A., Clark, C. T., Cooper, W. W. and Golany, B. (1985) A Developmental Study of Data Envelopment Analysis in Measuring the Efficiency of Maintenance Units in the U.S. Air Forces, *Annals of Operation Research*, 2, 95–112.

Charnes, A., Cooper, W. W., Lewin, A. Y. and Seiford, L. M. (1994) *Data Envelopment Analysis: Theory, Methodology and Application*, Boston, Dordrecht and London: Kluwer Academic Publishers.

Chen, T. (1999) Yard Operations in the Container Terminal – A Study in the 'Unproductive Moves', *Maritime Policy and Management*, 26(1), 27–38.

Chen, C. S., Lee, S. M. and Shen, Q. S. (1995) An Analytical Model for the Container Loading Problem, *European Journal of Operations Research*, 80(1), 68–76.

Chow, L. R. (1986) An Algorithm and Experimental Design for the Computer Control of Containers in the Port, *International Journal on Policy and Information*, 10(2), 31–42.

Chu, C. and Huang, W. (2002) Aggregates Cranes Handling Capacity of Container Terminals: The Port of Kaohsiung, *Maritime Policy and Management*, 29(4), 341–350.

Coeck, C., Haezendonck, E., Notteboom, T., Verbeke, A. and Winkelmans, W. (1997) The Competitiveness of Seaports: Business and Government Agenda's in Strategic Planning, in: *Essays in Honour and in Memory of Late Professor Emeritus of Maritime Economics Dr. Basil N. Metaxax*, Piraeus, 269–287.

Coelli, T. J. (1992) A Computer Program for Frontier Production Function Estimation: FRONTIER Version 2.0, *Economics Letters*, 39(1), 29–32.

Coelli, T., Prasada Rao, D. S. and Battese, G. E. (1998) *An Introduction to Efficiency and Productivity Analysis*, Boston, Dordrecht and London: Kluwer Academic Publishers.

Containerisation International Yearbook (1998) Quay Crane Build-up, *Containerisation International Yearbook 1998*, 16–17.

Cooper, W. W., Seiford, L. M. and Tone, K. (2000) *Data Envelopment Analysis: A Comprehensive Text with Models, Applications, References and DEA-Solver Software*, Boston: Kluwer Academic Publishers.

Coto-Millan, P., Banos-Pino, J. and Rodriguez-Alvarez, A. (2000) Economic Efficiency in Spanish Ports: Some Empirical Evidence, *Maritime Policy and Management*, 27(2), 169–174.

Cullinane, K. P. B. (2003a) *Port Competition in the Yangtze Delta: Plans and Prospects for the Future*, invited paper presented at the Seatrade London International Maritime Convention, 16–18 September, ExCel, London.

Cullinane, K. P. B. (2003b) The Productivity and Efficiency of Ports and Terminals: Methods and Applications, in Grammenos, C. T. (ed.), *The Handbook of Maritime Economics and Business*, Chapter 35, 803–831, LLP: London.

Cullinane, K. P. B. (2004) *Port Competition in China: The Case of Shanghai and Ningbo*, invited paper presented at the Critical Issues Facing the Port and Shipping Industry in the 21st Century Conference, 8 March, Singapore.

Cullinane, K. P. B. and Khanna, M. (1999) Economies of Scale in Large Containerships, *Journal of Transport Economics and Policy*, 33(2), 185–208.

Cullinane, K. P. B. and Khanna, M. (2000) Economies of Scale in Large Containerships: Optimal Size and Geographical Implications, *Journal of Transport Geography*, 8, 181–195.

Cullinane, K. P. B. and Song, D. W. (1998) Container Terminals in South Korea: Problems and Panaceas, *Maritime Policy and Management*, 25(1), 63–80.

Cullinane, K. P. B. and Song, D. W. (2001) The Administrative and Ownership Structure of Asian Container Ports, *International Journal of Maritime Economics*, 3(2), 175–197.

Cullinane, K. P. B. and Song, D. W. (2003) A Stochastic Frontier Model of the Productive Efficiency of Korean Container Terminals, *Applied Economics*, 35(3), 251–267.

Cullinane, K. P. B., Song, D. W. and Gray, R. (2002) A Stochastic Frontier Model of the Efficiency of Major Container Terminals in Asia: Assessing the Influence of Administrative and Ownership Structures, *Transportation Research A*, 36, 743–762.

Cullinane, K. P. B., Wang, T. and Cullinane, S. L. (2004) Container Terminal Development in Mainland China and its Impact on the Competitiveness of the Port of Hong Kong, *Transport Reviews*, 24(1), 33–56.

Daganzo, C. F. (1989) The Crane Scheduling Problem, *Transportation Research B*, 3, 159–175.

Daganzo, C. F. (1990) Crane Productivity and Ship delay in Ports, *Transportation Research Record*, 1251, 1–9.

Day, D. L., Lewin, A. Y. and Li, H. (1995) Strategic Leaders or Strategic Groups: A Longitudinal Data Envelopment Analysis of the U.S. Brewing Industry, *European Journal of Operational Research*, 80, 619–638.

De Borger, B., Kerstens, K. and Costa, A. (2002) Public Transit Performance: What does One Learn from Frontier Studies, *Transport Reviews*, 22(1), 1–38.

De Monie, G. (1987) Measuring and Evaluating Port Performance and Productivity, UNCTAD. *Monographs on Port Management No. 6 on Port Management* (Geneva, UNCTAD).

De Neufville, R. and Tsunokawa, K. (1981) Productivity and Returns to Scale of Container Ports, *Maritime Policy and Management*, 8(2), 121–129.

Deprins, D., Simar, L. and Tulkens, H. (1984) Measuring Labour-Efficiency in Post Offices, in Marchand, M., Pestieau, P. and Tulkens, H. (eds), *The Performance of Public Enterprises: Concepts and Measurement*, North-Holland Publishing Company, Amsterdam.

De Weille, J. and Ray, A. (1974) The Optimum Port Capacity, *Journal of Transport Economics and Policy*, VIII(3), 244–259.

Dowd, T. J. and Leschine, T. M. (1990) Container Terminal Productivity: A Perspective, *Maritime Policy and Management*, 17(2), 107–112.

Drewry Shipping Consultants (1998) *World Container Terminals: Global Growth and Private Profit*, London.

Dyson, R. G. (2000) Performance Measurement and Data Envelopment Analysis – Ranking are Ranks! *OR Insight*, 13(4), 3–8.

Edmond, E. D. and Maggs, R. P. (1976) Container Ship Turnaround Times at UK Ports, *Maritime Policy and Management*, 4, 3–19.

Edmond, E. D. and Maggs, R. P. (1978) How Useful are Queue Models in Port Investment Decisions for Container Berths? *Journal of the Operational Research Society*, 29(8), 741–750.

Engle, R. (1984) Wald, Likelihood Ratio and Lagrange Multiplier Tests in Econometrics, in Griliches, Z. and Intriligator, M. (eds), *Handbook of Econometrics*, Volume II, Amsterdam: Elsevier Science Publishers, 775–826.

Evers, J. J. M. and Koppers, S. A. J. (1996) Automated Guided Vehicle Traffic Control at a Container Terminal, *Transportation Research A*, 30(1), 21–34.

Farrell, M. J. (1957) The Measurement of Productive Efficiency, *Journal of Royal Statistical Society A*, 120, 253–281.

Fleming, D. K. (1997) World Container Port Rankings, *Maritime Policy and Management*, 24(2), 175–181.

Forsund, F. R. and Sarafoglou, N. (2002) On the Origins of Data Envelopment Analysis, *Journal of Productivity Analysis*, 17, 23–40.

Forsund, F. R., Lovell, C. and Schmidt, P. (1980) A Survey of Frontier Production Functions and of their Relationship to Efficiency Measurement, *Journal of Econometrics*, 13, 5–25.

Fossey, J. (1998) The Yangtse Pipeline, *Containerisation International*, July, 74.

Frankel, E. G. (1991) Port Performance and Productivity Measurement, *Ports and Harbours*, October, 11–13.

Fung, K. F. (2001) Competition between the Ports of Hong Kong and Singapore: A Structural Vector Error Correction Model to Forecast the Demand for Container Handling Services, *Maritime Policy and Management*, 28(1), 3–22.

Gallant, R. A. (1981) On the Bias in Flexible Functional Forms and an Essentially Unbiased Form: The Fourier Function, *Journal of Econometrics*, 15, 211–245.

Gehring, M. and Bortfeldt, A. (1997) A Genetic Algorithm for Solving Container Loading Problem, *International Transactions in Operational Research*, 4(5/6), 401–418.

Golany, B. and Tamir, E. (1995) Evaluating Efficiency-Effectiveness-Equality Trade-Offs – A Data Envelopment Analysis Approach, *Management Science*, 41(7), 1172–1184.

Gong, B. H. and Sickles, R. (1989) Finite Sample Evidence on the Performance of Stochastic Frontier Models Using Panel Data, *Journal of Productivity Analysis*, 1, 229–261.

Gong, B. H. and Sickles, R. (1992) Finite Sample Evidence on the Performance of Stochastic Frontiers and Data Envelopment Analysis Using Panel data, *Journal of Econometrics*, 51, 259–284.

Goss, R. O. (1982) Competition in Cargo Handling: Some Experience from Australia, *Maritime Policy and Management*, 9(1), 45–57.

Goss, R. O. (1990a) Economic Policies and Seaports – Part 1: The Economic Functions of Seaports, *Maritime Policy and Management*, 17(3), 207–219.

Goss, R. O. (1990b) Economic Policies and Seaports – Part 2: The Diversity of Port Policies, *Maritime Policy and Management*, 17(3), 221–224.

Goss, R. O. (1990c) Economic Policies and Seaports – Part 3: Are Port Authorities Necessary, *Maritime Policy and Management*, 17(4), 257–271.

Goss, R. O. (1990d) Economic Policies and Seaports – Part 4: Strategies for Port Authorities, *Maritime Policy and Management*, 17(4), 273–287.

Goulielmos, A. M. and Pardali, A. I. (2002) Container Ports in Mediterranean Sea: A Supply and Demand Analysis in the Age of Globalisation, *International Journal of Transport Economics*, 29(1), 91–117.

Greene, W. H. (1980) Maximum Likelihood Estimation of Econometric Frontier Functions, *Journal of Econometrics*, 13, 27–56.

Greene, W. H. (1990) A Gamma Distributed Stochastic Frontier Model, *Journal of Econometrics*, 46(1/2), 141–164.

Greene, W. H. (1993) The Econometric Approach to Efficiency Analysis, in Fried, H. O., Lovell, C. A. K. and Schmidt, S. S. (eds), *The Measurement of Productive Efficiency*, Oxford: Oxford University Press.

Greene, W. H. (2002) Fixed and Random Effects in Stochastic Frontier Models, Working paper, Department of Economics, Stern School of Business, New York University.

Grosskopf, S. (1993) Efficiency and Productivity, in Fried, H. O., Lovell, C. A. K. and Schmidt, S. S. (eds), *The Measurement of Productive Efficiency: Techniques and Applications*, Oxford: Oxford University Press, 160–194.

Guan, Y. and Cheung, R. K. (2004) The Berth Allocation Problem: Models and Solution Methods, *OR Spectrum*, 26, 75–92.

Hartman, T. E and Storbeck, J. E. (1996) Input Congestion in Loan Operations, *International Journal of Production Economics*, 46–47, 413–421.

Hausman, J. (1978) Specification Tests in Econometrics, *Econometrica*, 46, 1251–1271.

Hausman, J. and Taylor, W. (1981) Panel Data and Unobserved Individual effects, *Econometrica*, 49, 1377–1398.

Hayuth, Y. (1981) Containerisation and the Load Centre Concept, *Economic Geography*, 57, 160–176.

Heaver, T. (1995) The Implications of Increased Competition among Ports for Port Policy and Management, *Maritime Policy and Management*, 22(2), 125–133.

Heaver, T., Meersman, H., Moglia, F. and van de Voorde, E. (2000) Do Mergers and Alliances Influence European Shipping and Port Competition? *Maritime Policy and Management*, 27(4), 363–373.

Heaver, T., Meersman, H. and van de Voorde, E. (2001) Co-operation and Competition in International Container Transport: Strategies for Ports, *Maritime Policy and Management*, 28(3), 293–305.

Huybrechts, M., Meersman, H., van de Voorde, E., Van Hooydonck, E., Verbeke, A. and Winkelmans, W. (2002) *Port Competitiveness: an Economic and Legal Analysis of the Factors Determining the Competitiveness of Seaports*, Antwerpen, Editions De Boeck Ltd.

Imai, A., Nagaiwa, K. and Tat, C. W. (1997) Efficient Planning of Berth Allocation for Container Terminals in Asia, *Journal of Advanced Transportation*, 31(1), 75–94.

Imai, A., Nishimura, E. and Papadimitriou, S. (2001) The Dynamic Berth Allocation Problem for a Container Port, *Transportation Research B*, 35, 401–417.

Imai, A., Nishimura, E. and Papadimitriou, S. (2002a) Berth Allocation with Service Priority, *Transportation Research B*, 37, 437–457.

Imai, A., Nishimura, E., Papadimitriu, S. and Sasaki, K. (2002b) The Containership Loading Problem, *International Journal of Maritime Economics*, 4, 126–148.

Janelle, D. G. and Beuthe, M. (1997) Globalization and Research Issues in Transportation, *Journal of Transport Geography*, 5(3), 199–206.

Janelle, D. G. and Beuthe, M. (2002) Globalisation and Transportation: Contradictions and Challenges, in Black, W. R. and Nijkamp, P. (eds), *Social Change and Sustainable Transport*, Bloomington: Indiana University Press.

Jara-Díaz, S., Cortes, C., Vargas, A. and Martínez-Budría, E. (1997) Marginal Costs and Scale Economies in Spanish Ports, *25th European Transport Forum*, Proceedings Seminar L, PTRC, London, 137–147.

Jara-Díaz, S., Martinez-Budría, E., Cortes, C. and Basso, L. (2002) A Multioutput Cost Function for the Services of Spanish Ports' Infrastructure, *Transportation*, 29(4), 419–437.

Jondrow, J., Lovell, C., Materov, I. and Schmidt, P. (1982) On the Estimation of Technical Inefficiency in the Stochastic Frontier Production Model, *Journal of Econometrics*, 19, 233–238.

Jorde, T. M. and Teece, D. J. (1989) Competition and Co-operation: Striking the Right Balance, *California Management Review*, 31(3), 25–37.

Kalirajan, K. P. and Shand, R. T. (1999) Frontier Production Functions and Technical Efficiency Measures, *Journal of Economic Surveys*, 13(2), 149–172.

Kim, K. H. (1997) Evaluation of the Number of Rehandles in Container Yards, *Computers and Industrial Engineering*, 32(4), 701–711.

Kim, K. H. and Bae, J. W. (1998) Re-marshaling Export Containers in Port Container Terminals, *Computers and Industrial Engineering*, 35(3/4), 655–658.

Kim, K. H. and Kim, H. B. (1998) The Optimal Determination of the Space Requirement and the Number of Transfer Cranes for Import Containers, *Computers and Industrial Engineering*, 35(3/4), 427–430.

Kim, K. H. and Kim, K. Y. (1999) An Optimal Routing Algorithm for a Transfer Crane in Port Container Terminals, *Transportation Science*, 33(1), 17–33.

Kim, M. and Sachish, A. (1986) The Structure of Production, Technical Change and Productivity in a Port, *Journal of Industrial Economics*, 35(2), 209–223.

Kohersa, T., Huang, M. and Kohers, N. (2000) Market Perception of Efficiency in Bank Holding Company Mergers: The Roles of the DEA and SFA Models in Capturing Merger Potential, *Review of Financial Economics*, 9, 101–120.

Kopp, R. J., Smith, V. K. and Vaughan, W. J. (1982) Stochastic Cost Frontiers and Perceived Technical Inefficiency, in Smith, V. K. (ed.), *Advances in Applied Microeconomics*, 2, Greenwich, CN: JAI Press.

Kozan, E. (1997) Comparison of Analytical and Simulation Planning Models of Seaport Container Terminals, *Transportation Planning and Technology*, 20, 235–248.

Kumar, S. and Hoffmann, J. (2002) Globalisation: The Maritime Nexus, in Grammenos, C. T. (ed.), *The Handbook of Maritime Economics and Business*, London: LLP.

Kumbhakar, S. C. and Lovell, C. (2000) *Stochastic Frontier Analysis*, Cambridge: Cambridge University Press.

Lai, K. K. and Lam, K. (1994) A Study of Container Yard Equipment Allocation Strategy in Hong Kong, *International Journal of Modeling and Simulation*, 14(3), 134–138.

Lansink, A. O., Silva, E. and Stefanou, S. (2001) Inter-firm and Intra-firm Efficiency Measures, *Journal of Productivity Analysis*, 15, 185–199.

Leibenstein, H. (1966) Allocative Efficiency *vs.* 'X-efficiency', *American Economic Review*, 56, 392–415.

Leibenstein, H. (1973) Competition and X-Efficiency: Reply, *Journal of Political Economy*, 81(3), 765–777.

Liu, Z. (1992) *Ownership and Productive Efficiency: with Reference to British Ports*, PhD Thesis, Queen Mary and Westfield College, University of London.

Liu, Z. (1995) The Comparative Performance of Public and Private Enterprises: The Case of British Ports, *Journal of Transport Economics and Policy*, 29(3), 263–274.

Lovell, C. (1993) Production Frontiers and Productive Efficiency, in Fried, H. O., Lovell, C. A. K. and Schmidt, S. S. (eds), *The Measurement of Productive Efficiency: Techniques and Applications*, Oxford: Oxford University Press, 3–67.

Lovell, C. and van den Beckaut, P. (1993) Frontier Tales: DEA and FDH, in *Mathematical Modelling in Economics: Essays in Honor of Wolfgang Eichhorn*, Berlin, New York: Springer-Verlag, 446–457.

Marlow, P. B. and Paixão, A. C. (2001) Agility, a Key Enabler in Port Competition, *Proceedings of The International Association of Maritime Economists Annual Conference 2001*, 18–20 July, Hong-Kong, 102–114.

Marlow, P. B. and Paixão, A. C. (2002) Measuring Lean Ports Performance, *Proceedings of the International Association of Maritime Economists Conference,* November, Panama, 13–15.

Martinez-Budria, E., Diaz-Armas, R., Navarro-Ibanez, M. and Ravelo-Mesa, T. (1999) A study of the Efficiency of Spanish Port Authorities Using Data Envelopment Analysis, *International Journal of Transport Economics,* XXVI(2), 237–253.

McKenzie, C. R. and Takaoka, S. (2003) 2002: A LIMDEP Odyssey, *Journal of Applied Econometrics,* 18(2), 241–247.

McLellan, R. G. (1997) How Big is Too Big? *Maritime Policy and Management,* 24(2), 193–211.

Meersman, H. and van de Voorde, E. (2002) Port Management, Operation and Competition: A Focus on North Europe, in Grammenos, C. T. (ed.), *The Handbook of Maritime Economics and Business,* LLP: London.

Meeusen, W. and van den Broeck, J. (1977) Efficiency Estimation from Cobb–Douglas Production Functions with Composed Error, *International Economic Review,* 18(2), 435–444.

Murillo-Zamorano, L. R. and Vega-Cervera, J. A. (2001) The Use of Parametric and Non-Parametric Frontier Methods to Measure the Productive Efficiency in the Industrial Sector: A Comparative Study, *International Journal of Production Economics,* 69, 265–275.

Musso, E., Ferrari, C. and Benacchio, M. (1999) On the Global Optimum Size of Port Terminals, *International Journal of Transport Economics,* 26(3), 415–437.

Nanuenberg, E., Andrijuk, Y. and Eisinger, M. (2001) Reconsideration of Discharge Data to Measure Competition in the Hospital Industry, *Health Economics,* 10(3), 271–276.

Noritake, M. and Kimura, S. (1983) Optimum Number and Capacity of Seaport berths, *Journal of Waterway, Port, Coastal and Ocean Engineering,* 109(33), 323–339.

Noritake, M. and Kimura, S. (1990) Optimum Allocation and Size of Seaports, *Journal of Waterway, Port, Coastal and Ocean Engineering,* 116(2), 287–299.

Norman, M. and Stoker, B. (1991) *Data Envelopment Analysis, the Assessment of Performance,* John Wiley & Sons Ltd, Chichester, England.

Notteboom, T., Coeck, C. and van den Broeck, J. (2000) Measuring and Explaining the Relative Efficiency of Container Terminals by Means of Bayesian Stochastic Frontier Models, *International Journal of Maritime Economics,* 2(2), 83–106.

Notteboom, T. E. and Winkelmans, W. (2001) Structural Changes in Logistics: How Will Port Authorities Face the Challenge?, *Maritime Policy and Management,* 28(1), 71–89.

Notteboom, T., Coeck, C., Verbeke, A. and Winkelmans, W. (1997) Containerisation and the Competitive Potential of Upstream Urban Ports in Europe, a Comment, *Maritime Policy and Management,* 24(3), 285–289.

Odeck, J. (2001) Comparison of Data Envelopment Analysis and Deterministic Parametric Frontier Approaches: An Application in the Norwegian Road Construction Sector, *Transportation Planning and Technology,* 24(2), 111–134.

Ondrich, J. and Ruggiero, J. (2001) Efficiency Measurement in the Stochastic Frontier Model, *European Journal of Operational Research,* 129(2), 434–442.

Oum, T. H., Waters, W. G. and Yu, C. Y. (1999) A Survey of Productivity and Efficiency Measurement in Rail Transport, *Journal of Transport Economics and Policy,* 33(1), 9–42.

Peterkofsky, R. I. and Daganzo, C. F. (1990) A Branch and Bound Method for the Crane Scheduling Problem, *Transportation Research B*, 24(3), 159–172.

Peters, H. J. F. (2001) Developments in Global Seatrade and Container Shipping Markets: Their Effects on the Port Industry and Private Sector Involvement, *International Journal of Maritime Economics*, 3(1), 3–26.

Ray, S. C. (2002) William W. Cooper: A Legend in His Own Times, *Journal of Productivity Analysis*, 17, 7–12.

Reinhard, S., Lovell, C. and Thijssen, G. J. (2000) Environmental Efficiency with Multiple Environmentally Detrimental Variables; Estimated with SFA and DEA, *European Journal of Operational Research*, 121(2), 287–303.

Rhoades, S. A. (1993) The Herfindahl–Hirschman Index, *Federal Reserve Bulletin*, March, 188–189.

Riccardo, C. (2000) A Generalized Measure of Competition, *Applied Economics Letters*, 7(7), 479–481.

Richmond, J. (1974) Estimating the Efficiency of Production, *International Economic Review*, 15(2), 515–521.

Robinson, R. (1978) Size of Vessels and Turnround Times, *Journal of Transport Economics and Policy*, 12(2), 161–178.

Robinson, D. (1999) *Measurements of Port Productivity and Container Terminal Design: A Cargo Systems Report*, London: IIR Publications.

Robinson, R. (2002) Ports as Elements in Value-Driven Chain Systems: The New Paradigm, *Maritime Policy and Management*, 29(3), 241–255.

Robinson, R. and Reyes, J. M. (1998) ASEAN and Australian Ports: Some Aspects of Efficiency and Productivity in the Early 1980s, in Trace, K. *et al.* (eds), *Handmaiden of Trade*, Singapore University Press, 113–144.

Roll, Y. and Hayuth, Y. (1993) Port Performance Comparison Applying Data Envelopment Analysis (DEA) *Maritime Policy and Management*, 20(2), 153–161.

Ross, A. and Droge, C. (2002) An Integrated Benchmarking Approach to Distribution Center Performance Using DEA Modeling, *Journal of Operations Management*, 20, 19–32.

Sarafoglou, N. (1998) The Most Influential DEA Publications: A Comment on Seiford, *Journal of Productivity Analysis*, 9, 279–281.

Saundry, R. and Turnbull, T. (1997) Private Profit, Public Loss: The Financial and Economic Performance of U.K. Ports, *Maritime Policy and Management*, 24(4), 319–334.

Scherer, F. M. and Ross, D. (1990) *Industrial Market Structure and Economic Performance*, 3rd edn, Boston, MA: Houghton Mifflin.

Schmidt, P. (1976) On the Statistical Estimation of Parametric Frontier Production Functions, *Review of Economics and Statistics*, 58(2), 238–239.

Schmidt, P. and Sickles, R. C. (1984) Production Frontiers and Panel Data, *Journal of Business and Economic Statistics*, 2(4), 367–374.

Schonfeld, P. and Frank, S. (1984) Optimising the Use of a Containership Berth, *Transportation Research Record*, 984, 56–62.

Schonfeld, P., Asce, A. M. and Sharafeldien, O. (1985) Optimal Berth and Crane Combinations in Container Ports. *Journal of Waterway, Port, Coastal and Ocean Engineering*, ASCE, 111(6), 1060–1072.

Schwartzman, D. (1973) Competition and Efficiency: Comment, *Journal of Political Economy*, 81(3), 756–764.

Seiford, L. M. (1996) Data Envelopment Analysis: The Evolution of the State of the Art (1978–1995), *Journal of Productivity Analysis*, 7, 99–137.

Seiford, L. M. and Thrall, R. (1990) Recent Development in DEA: The Mathematical Programming Approach to Frontier Analysis, *Journal of Econometrics*, 46 (1/2), 7–38.

Sharma, K., Leung, P. and Zaleski, H. M. (1997) Productive Efficiency of the Swine Industry in Hawaii: Stochastic Frontier vs. Data Envelopment Analysis, *Journal of Productivity Analysis*, 8, 447–459.

Sherman, H. D. (1981) *Measurement of Hospital Technical Efficiency: A Comparative Evaluation of Data Envelopment Analysis and Other Techniques for Measuring and Locating Efficiency in Health Care Organisations*, PhD Thesis, Boston: Graduate School of Business, Harvard University.

Slack, B. (1993) Pawns in the Game: Ports in a Global Transportation System. *Growth and Change*, 24, 579–588.

Song, D. W. (2002) Regional Container Competition and Co-operation: The Case of Hong Kong and South China, *Journal of Transport Geography*, 10(2), 99–110.

Song, D. W. (2003) Port Co-opetition in Concept and Practice, *Maritime Policy and Management*, 30(1), 29–44.

Song, D. W., Cullinane, K. P. B. and Roe, M. (2001) *The Productive Efficiency of Container Terminals, An Application to Korea and the UK*, Ashgate: Aldershot, England.

Stevenson, R. E. (1980) Likelihood Functions for Generalised Stochastic Frontier Functions Estimation, *Journal of Econometrics*, 13(1), 57–66.

Stigler, G. J. (1976) The Existence of X-Efficiency, *American Economic Review*, 66(1), 213–216.

Stolp, C. (1990) Strengths and Weaknesses of Data Envelopment Analysis: An Urban and Regional Perspective, *Computers, Environment and Urban Systems*, 14(2), 103–116.

Sun, S. (2002) Measuring the Relative Efficiency of Police Precincts Using Data Envelopment Analysis, *Socio-Economic Planning Sciences*, 36, 51–71.

Suykens, F. (1983) A Few Observations on Productivity in Seaports, *Maritime Policy and Management*, 10(1), 17–40.

Suykens, F. (1986) Ports Should be Efficient (Even When this Means that Some of them are Subsidized) *Maritime Policy and Management*, 13(2), 105–126.

Suykens, F. and van de Voorde (1998) A Quarter of a Century of Port Management in Europe: Objectives and Tools, *Maritime Policy and Management*, 25(3), 251–261.

Tabernacle, J. B. (1995) A Study of the Changes in Performance of Quayside Container Cranes, *Maritime Policy and Management*, 22(2), 115–124.

Taleb-Ibraimi, M., De Castilho, B. and Daganzo, C. F. (1993) Storage Space vs. Handling Work in Container Terminals, *Transportation Research B*, 27, 13–32.

Talley, W. K. (1994) Performance Indicators and Port Performance Evaluation. *The Logistics and Transportation Review*, 30(4), 339–352.

Talley, W. K. (1998) Optimum Throughput and Performance Evaluation of Marine Terminals. *Maritime Policy and Management*, 15(4), 327–331.

Talley, W. K. (2000) Ocean Container Shipping: Impacts of a Technological Improvement, *Journal of Economic Issues*, 34(4), 933–948.

Thanassoulis, E. (2001) *Introduction to Theory and Application of Data Envelopment Analysis*, Norwell, MA: Kluwer Academic Publishers.

Tongzon, J. L. (1993) The Port of Melbourne Authority's Pricing Policy: Its Efficiency and Distribution Implications, *Maritime Policy and Management*, 20(3), 197–203.

Tongzon, J. (1995) Systematising International Benchmarking for Ports, *Maritime Policy and Management*, 22(2), 171–177.

Tongzon, J. (2001) Efficiency Measurement of Selected Australian and Other International Ports Using Data Envelopment Analysis, *Transportation Research Part A: Policy and Practice*, 35(2), 113–128.

Tongzon, J. and Ganesalingam, S. (1994) An Evaluation of ASEAN Port Performance and Efficiency, *Asian Economic Journal*, 8(3), 317–330.

Tulkens, H. (1993) On FDH Efficiency Analysis: Some Methodological Issues and Applications to Retail Banking, Courts and Urban Transit, *Journal of Productivity Analysis*, 4, 183–210.

Tulkens, H. and van den Eeckaut, P. (1995) Nonparametric Efficiency, Progress and Regress Measures for Panel-Data: Methodological Aspects, *European Journal of Operational Research*, 80(3), 474–499.

Turnbull, P. and Weston, S. (1993) Co-operation or Control? Capital Restructuring and Labour Relations on the Docks, *British Journal of Industrial Relations*, 31(1), 115–34.

UNCTAD (1976) *Port Performance Indicators*, TD/B/ C.4/131/Supp.1/Rev.1, New York, US: United Nations Conference on Trade and Development.

UNCTAD (1992) *Port Marketing and the Challenge of the Third Generation Port*, TD/B/C. 4/AC. 7/14, Geneva.

UNCTAD (1995) *Comparative Analysis of Deregulation, Commercialisation and Privatisation of Ports*, UNCTAD/SDD/PORT/3, Geneva.

UNCTAD (2001) *Review of Maritime Transport*, Geneva: United Nations.

Valentine, V. F. and Gray, R. (2001) The Measurement of Port Efficiency Using Data Envelopment Analysis, *Proceedings of the 9th World Conference on Transport Research*, 22–27 July, Seoul, South Korea.

Vancil, R. F. (1973) What Kind of Management Tool Do You Need? *Harvard Business Review*, 51(2), 75–86.

van den Broeck, J., Koop, G., Osiewalski, J. and Steel, M. F. J. (1994) Stochastic Frontier Models: a Bayesian Perspective, *Journal of Econometrics*, 61, 273–303.

van den Eeckaut, P., Tulkens, H. and Jamar, M. A.(1993) Cost Efficiency in Belgian Municipalities, in Fried, H. O., Lovell, C. A. K. and Schmidt, S. S. (eds), *The Measurement of Productive Efficiency*, New York: Oxford University Press.

Verhoeff, J. M. (1981) Seaport Competition: Some Fundamental and Political Aspects, *Maritime Policy and Management*, 8(1), 49–60.

Vickers, J. and Yarrow, G. (1988) *Privatisation: An Economic Analysis*, Cambridge, MA: The MIT Press.

Waldman, D. (1982) A Stationary Point for the Stochastic Frontier Likelihood, *Journal of Econometrics*, 18, 275–279.

Wang, T., Song, D. W. and Cullinane, K. P. B. (2002) The Applicability of Data Envelopment Analysis to Efficiency Measurement of Container Ports, *Proceedings of the International Association of Maritime Economists Conference*, Panama, 13–15 November.

Wilson, I. D. and Roach, P. (2000) Container Stowage Planning: A Methodology for Generating Computerised Solutions, *Journal of the Operational Research Society*, 51(11), 1248–1255.

Winkelmans, W. (1998) Strategic Seaport Planning in Search for Better Use of Core Competencies and Core Competitive Advantages, in Misztal, K. and Zurek, J. (eds), *Maritime Transport and Economic Reconstruction*, Institute of Maritime Transport and Seaborne Trade, University of Gdànsk, Gdànsk, 25–35.

Winkelmans, W. (2002) Strategic Seaport Planning: In Search of Core Competency and Competitive Advantage, *Ports and Harbors*, 47(3), 17–21.

Index

Access, 19, 28, 58, 59
Additive Model, 42, 43, 67, 68, 74
Allocative Efficiency, 4–7, 73, 145
ANOVA, 94, 95, 101, 102, 105, 108, 110, 113
Antwerp, 24, 95, 124, 130, 136, 137
Asia, 23, 24, 69, 70, 72, 86
Asset(s), 83, 87, 141
Assumptions, 10, 21, 36–8, 49–53, 56–61, 72–5, 87, 89, 100, 101, 106, 107, 110, 111, 114, 144
Australia, 10, 22, 30, 64, 67, 68, 86

Bayesian, 69, 70
BCC Model, 42, 43, 50, 68, 94–7, 103–5, 122, 136
Behavioural, 3, 5, 6, 10
Benchmark, 69, 72, 73, 88, 95
Benchmarking, 4, 9, 25, 31, 32, 65, 66, 68, 143
Berth, 10, 16, 17, 28, 63, 64, 66–8, 83–5, 92
Best Practice, 48, 49, 54, 98, 114, 117
Bias, 85, 128
Book Value, 84

Capacity, 17, 19, 23–6, 30, 84, 85, 88, 110
Capital, 2, 5, 15, 24, 64, 67, 69, 70, 72, 77, 84, 85
Cargo, 1, 10, 14, 15, 16, 18, 20, 22–4, 63, 64, 67, 70, 80–4, 92
Cargo Handling, 10, 22, 23, 63, 64, 70, 83–5, 92, 106
Carriers, 16, 19, 85, 89, 92
CCR Model, 39, 41–3, 50, 67, 68, 74, 94–7, 103–5, 110, 113, 118–22, 127, 130, 136, 137
Central Government, 11, 25, 26
Centralisation, 24, 26
China, 22, 24, 87, 90, 91, 124, 126, 129, 131, 134, 135
Cluster Analysis, 66

Cobb-Douglas, 52, 56, 60, 70, 71, 74, 83, 89, 92, 114, 143, 145
Coefficients, 60, 89, 94, 99, 100, 101, 108, 109, 110, 111
Corrected Ordinary Least Squares (COLS), 54, 93, 98, 100, 101–3, 113
Competition, 8–13, 20–32, 71–2, 81, 119–30, 139, 141–3
Competitive, 5, 8, 9, 10, 16, 18, 21–7, 31, 59, 72, 88, 119–22, 146
Competitiveness, 20, 25, 32, 118, 145, 146
Comprehensive Port, 27
Concentration of Industry, 119–22, 124, 130
Constant Returns to Scale, 5, 42, 50, 63, 73, 74, 94, 95
Construction, 26, 45, 46
Container(s), 1, 2, 8–26, 31–3, 37, 40, 48, 63–76, 80–95, 98, 103–4, 106, 108, 110, 112–22, 126, 129, 131, 136, 138–46
Container Port, 8–11, 13, 17, 18, 22, 23, 25, 32, 33, 67, 68, 74, 81, 87, 90–2, 98, 106, 112, 114, 116–21, 126, 139–46
Container Terminals, 72, 85, 86, 91, 136
Container Yard, 16, 17, 85, 106
Containerisation, 14, 20, 22, 23, 65, 87, 91
Containership, 16, 17, 20
Contemporaneous Analysis, 45, 47, 48, 103–5, 113, 114, 117–19, 122, 127, 130, 144
Contestable market(s), 27, 31
Correlation, 83, 93–5, 101–3, 106, 108, 110, 113, 117–19, 122, 130
Cost(s), 1, 4, 5, 7, 10, 17, 18, 20, 25, 26, 30, 33, 36, 44, 47, 48, 59, 63, 67, 70, 80, 82, 84, 88
Cost Frontier, 4, 33, 47

Countr(y)ies, 9, 22–6, 31, 69, 80, 124

Cross-Section(al), 10, 34, 37–40, 45, 48, 51, 58–60, 65, 72, 74, 89–93, 102, 103, 113, 116, 117, 122, 136, 141–3

Customers, 8, 14, 20, 21, 25, 26, 79, 80, 82

Dalian, 24, 124, 130, 131

Data, 10–12, 34, 36–51, 54, 58–61, 63, 65, 68–75, 77, 79, 82, 85–93, 98, 99, 102, 103, 106–10, 113–19, 122, 127, 129, 136, 141–4

Data Collection, 12, 77, 142

Data Envelopment Analysis (DEA), 10, 11, 36–46, 49–52, 63, 66–8, 72–7, 84, 87, 92–8, 103–5, 113–15, 118, 119, 122, 127, 130, 136, 137, 142–5

Decentralisation, 24, 26

Decision Making Unit(s) (DMUs), 2, 6, 40, 73

Decreasing Returns To Scale, 63

Degrees of Freedom, 99, 100

Deregulation, 21, 59, 72

Deterministic, 36, 49, 51, 54–6, 75, 98, 101–2

Development, 1, 4, 14, 15, 18, 23, 28, 34, 40, 42, 44, 59, 61, 64, 65, 81, 87, 91, 104, 112, 117

Distribution, 15, 19, 20, 48, 54, 55, 57, 75, 99, 108, 113, 127, 136, 144

Distributional, 58, 59, 72, 100, 101

Dock Workers, 86, 87

Dummy Variable, 61, 106

Econometric, 36, 38, 51, 53, 59, 60, 61, 63, 100, 143, 144

Economic(s), 1, 2, 4, 6, 9–13, 18, 20, 21, 26–32, 42, 45, 59–73, 79–80, 85, 88, 91, 98, 115, 116, 120, 122, 136, 140–2, 145

Economies of Scale, 1, 3–5, 10, 18, 20, 21, 26, 31, 42, 43, 50, 63, 69, 71–4, 83, 94–7, 103, 106, 117–19, 138–42

Economists, 10, 21, 24, 25, 31

Effectiveness, 64, 65, 76, 82, 83, 92

Efficient, 3, 5–7, 18–21, 23, 27, 31–3, 37, 40, 43–9, 52, 54, 65, 68–71, 74, 78, 83, 85, 94–8, 112, 113, 119, 137, 145

Empirical, 9–12, 18, 30, 61, 76, 79, 93, 98, 103, 111, 113–15, 118, 136, 138, 140–3

Employee(s), 82–5

Equation(s), 37, 38, 42, 50, 52, 53, 55, 56, 60, 89, 99, 120, 121

Equipment, 8, 19, 20, 25, 55, 64, 70, 80–7, 110, 118

Error(s), 55–9, 100

Estimates, 11, 12, 37, 43, 48, 53–8, 63, 68, 69, 73–5, 85, 92, 94, 95, 98–119, 122, 136, 142–5

Estimation, 11, 12, 36, 37, 51, 53–63, 69, 73, 75, 89, 92, 93, 98–115, 136, 140, 142

Europe(an), 14, 21, 24, 64, 68–71, 122, 124, 137

Exogenous, 44, 53, 89

Exponential, 36, 55, 57, 100, 108

Factor(s), 4, 10, 44, 48, 63, 66, 69, 71, 83, 88, 145, 146

Free Disposal Hull (FDH), 11, 36, 37, 49–52, 75, 93–8, 102, 103, 113, 118, 122, 130, 136, 137, 144

Feeder, 71

Firm(s), 2, 4–10, 21, 27, 31–4, 37–61, 65, 73–9, 94–7, 101–5, 110, 111, 114, 117, 120–2, 136, 142, 145

Fixed Effects, 11, 61, 63, 106–9

Fordism, 18

Fork Lifts, 84, 85

Forwarders, 16

Fractional Programming (FP), 41, 42

Freight, 1, 19

Frontier, 4–8, 33–7, 40–3, 47, 48, 50–61, 63, 66, 69, 70–5, 89, 93, 98–102, 107–13, 122, 141–5

Functional Form, 36, 49, 52, 56

Gamma Distribution, 57, 100, 101, 103

Gantry Crane(s), 66, 70, 85, 86, 89, 92

Gateway, 14, 22
Generalised Least Squares (GLS), 61, 106–10
Geographic, 22, 23, 120, 122
Germany, 87, 124
Gioia Tauro, 84, 85, 90, 98, 124
Globalisation, 1, 9, 18, 19
Goodness of Fit, 89
Goods, 2, 14, 18, 22, 28

Half-Normal Distribution, 36, 55, 100, 103
Hamburg, 23, 24, 69–71, 95, 97, 124, 130, 137
Hausman Test Statistic, 107
Herfindahl-Hirschman Index (HHI), 119–22, 124, 130
Hinterland(s), 8, 9, 18, 20, 22–5, 120, 122
Histogram, 99, 107
Hutchinson International Terminals (HIT), 69, 124, 137
Hong Kong, 10, 17, 22, 69, 95, 104, 112, 124, 130, 136, 137
Hub, 20, 71, 88, 122
Hybrid Port, 27
Hypothes(is)es, 4, 12–15, 18, 21, 27, 31, 32, 69, 70, 100, 107, 116–22, 126, 128, 136, 141, 142

Incentive(s), 30, 97
Incumbent(s), 120
Industrial Organisation, 11, 21, 29, 32, 141
Inefficien(t)cy, 3, 7, 8, 10, 33, 34, 40, 43, 45, 48, 51, 52, 55–60, 68, 71–5, 79, 89, 95, 97–101, 105, 113, 118, 122, 143, 146
Inference(s), 12, 53, 63, 117, 136
Information, 3, 5, 6, 9, 15–19, 25, 43, 44, 52, 58, 59, 65, 74, 78–82, 86, 87, 93, 94, 98, 136
Infrastructure, 14, 26–8, 106, 110
Innovation(s), 7, 8, 24, 26, 58, 103, 114
Input(s), 2–9, 12, 15, 34, 37–52, 58, 60, 65–9, 73, 75, 77, 79–89, 92, 106, 142, 145
Input-Oriented, 6, 87, 88

Instrumental Variable, 61, 106
Intercept, 54, 60, 61
Intermodal(ism), 14, 23
International, 1, 9, 14, 23, 25, 67, 68, 88, 140
Inter-Port Competition, 22–7, 121, 122
Intertemporal Analysis, 46–8, 103–5, 110, 114, 117–19, 122, 127, 130, 144
Intra-Port Competition, 2, 13, 27, 29, 30, 32, 141
Investment, 8, 11, 20, 25, 26, 28, 48, 82, 84, 110, 118
Isoquant, 6, 33, 34
Italy, 71, 124

Joint Venture(s), 86, 130

Kaohsiung, 69, 124, 130
Keelung, 98, 124, 130
Klang, 95, 98, 124, 130, 137
Kurtosis, 90, 91, 94, 101

Labour, 2, 5, 14, 15, 19, 77, 80–7
Lagrange Multiplier, 107
Land, 2, 14, 16, 17, 28, 83, 84
Landbridge, 23
Landlord Port(s), 27–9
Landowner Function, 29
Le Havre, 23, 24, 69–71
Leanness, 88
Least Squares, 61, 89, 106, 107
Likelihood Ratio, 99
LIMDEP, 59, 100
Linear Programming (LP), 34, 36, 41–4, 49, 50, 53, 55, 56, 60, 89, 92, 113
Liner Compan(y)ies, 1, 8, 9, 16, 17, 19, 20, 25, 40, 64, 80, 88, 131, 134, 135
Local Authorit(y)ies, 25–6
Logarithmic, 38, 53, 89
Logistics, 8, 14, 15, 18, 19
Long Beach, 24, 124, 130
Longitudinal, 10, 58
Los Angeles, 24, 84, 95, 130
Lumpiness, 25, 26, 30

Maersk-Sealand, 88
Malaysia, 22, 88, 90, 124
Management, 2, 8, 9, 10, 13, 17–19,
  22, 28–30, 64, 71, 80, 81, 88, 97,
  104, 112, 117, 140, 144
Management Science, 2, 39
Managerial Implications, 13, 25, 31,
  32, 73, 141, 143
Manila, 84, 85, 95, 124, 130, 137
Market(s), 8–10, 13, 18–22, 24–7, 30,
  31, 59, 72, 80, 84, 88, 118–22,
  141, 143
Market Concentration, 119–22,
  124, 130
Market Share, 24, 120
Market Structure(s), 8–10, 13, 18–22,
  24, 31
Marketing, 19, 26, 28
Mathematical Programming, 34, 36,
  41–4, 49, 50, 53, 55, 56, 60, 89,
  92, 113
Matri(x)ces, 29, 37, 38, 50, 72, 126
Maximisation, 3, 5, 80, 81
Maximum Likelihood Estimation
  (MLE), 53, 54, 56, 58, 61, 98–100,
  106–10, 113
Mean, 8, 27, 55, 90, 91, 94, 101, 104,
  119, 136
Measurement, 3, 11, 33–7, 49, 50,
  56, 58, 61, 63, 66, 69, 72, 75, 77,
  80, 82–4, 86, 140
Measuring Efficiency, 4, 8, 11,
  141, 144
Median, 90, 91, 94, 101
Mediterranean, 24, 69, 70
Melbourne, 86
Merger, 20, 26
Minimisation, 5, 65, 80, 81
Mobile Cranes, 85, 106
Modal Choice, 19
Model(s), 3, 6, 29, 33, 36–44, 47–63,
  66, 68–75, 77, 79, 82, 84–9,
  92–103, 106–14, 116, 118, 119,
  122, 127, 130, 136, 138, 141–5
Model Orientation, 77
Model Specification, 74, 77, 110
Modes, 1, 19, 20, 66, 89
Modified Ordinary Least Squares
  (MOLS), 54, 55

Monopolistic, 10, 21, 22, 24, 30
Monopoly, 10, 18, 21, 23, 30, 32,
  80, 121
Modern Teminals Limited (MTL), 69,
  124, 137

Nanjing, 98, 124, 130
National, 9, 22–6, 31, 88, 143
Network(s), 18, 19, 26
Ningbo, 24, 124
Node, 15, 19
Noise, 43, 56, 58, 60, 89, 145
Non-Parametric, 36–9, 44, 48, 49, 51,
  75, 76, 93, 102, 110, 113–16, 138,
  141–4
Normality, 55
Null Hypothesis, 100, 107

Objectives, 6, 11, 31, 39, 53, 65,
  77–81, 92, 140, 142, 146
Observations, 34, 36, 37, 45–9, 54, 55,
  58, 59, 67, 68, 70, 90, 95, 104, 128
Oligopolistic, 10, 22, 30
Oligopoly, 23, 30
Operations, 3, 9, 13, 16, 17,
  82–4, 140
Operator Function, 28, 29
Optimise, 6, 9, 20, 42, 142
Optimum, 10, 16, 17, 63, 145
Ordinary Least Squares (OLS), 36, 54,
  55, 98–100, 107–9
Output(s), 2–7, 12, 33–45, 47, 49–52,
  54, 55, 60, 65, 66, 68, 69, 73, 75,
  77–83, 87–9, 92, 120, 142, 145
Output Variable(s), 38, 51, 68, 73, 79,
  80, 82, 88, 92, 142
Output-Oriented, 88, 92
Outsourcing, 19
Overcapacity, 20, 25, 118
Ownership, 10, 11, 27–31, 67–70, 72,
  126, 127, 129–35, 140–3

Pacific, 23, 24
Panama Canal, 23
Panel Data, 11, 37, 38, 45–8, 51,
  58–61, 65, 69, 72, 74, 75, 89, 90,
  92, 93, 98, 103, 106, 107, 109,
  110, 113, 114, 117, 122, 136,
  141, 143, 144

Parameters, 36, 42, 44, 53–5
Parametric, 36–8, 49, 51, 52, 75–7, 89, 92, 93, 98–102, 110–14, 116, 138, 141–5
Passengers, 14, 28
Pearson Coefficient(s), 93, 117–19, 122, 130
Performance, 2–4, 7–10, 13, 18, 20, 24–32, 37, 43, 44, 63–7, 72–4, 77, 78, 80, 82–4, 101, 104, 139, 141, 145
Philadelphia, 18
Piecewise, 33, 34, 36, 42, 49, 50
Polic(y)ies, 4, 6, 8, 10, 11, 13, 21–6, 28–32, 48, 68, 70, 72, 73, 83, 85, 88, 122, 141–4, 146
Port(s), 1, 2, 8–33, 37, 40, 48, 63–77, 79–95, 97, 98, 101, 103–8, 110–22, 126–31, 135, 136, 138–46
Port Authorit(y)ies, 9, 27–9, 31, 64, 83
Port Charges, 10, 30, 83
Port Facilities, 20, 27, 28, 30
Port Function Matrix, 28, 72, 126
Port Industry, 8, 9, 18, 23–5, 29, 31, 32, 37, 48, 64, 66, 75, 76, 80, 87, 103, 121, 141, 144, 145
Port of Tanjung Pelepas (PTP), 88, 90
Port Operators, 9, 30
Port Performance, 64, 66
Port Planning, 9, 87
Port Pricing, 146
Port Production, 13, 32, 66, 68, 74, 83
Port Users, 13, 18, 83
Prices, 5, 7, 44, 80, 145
Principal Components, 66
Private Sector, 9, 10, 27–31, 69, 70, 72, 86, 126
Privatisation, 9, 10, 25, 29–31, 69, 70, 72, 73, 139, 141, 143
Produce, 2, 6, 11, 33, 34, 37, 45, 47, 50, 52, 53
Producer(s), 1, 5, 18, 37, 54, 55, 58–60
Product(s), 1, 2, 18, 79, 87, 120
Production, 1–8, 10–16, 18–21, 31–4, 36, 39, 40, 42, 43, 46–56, 58–61,

63, 65, 66, 68–70, 73, 74, 79–85, 87, 88, 91–5, 97–9, 103, 106–19, 136, 138, 140–5
Production Frontier, 4, 6, 7, 33, 40, 42, 43, 51, 52, 54, 74, 98, 114
Production Function, 52, 74, 85, 114, 143–5
Production Process, 7, 63, 84
Productivity, 2–8, 10, 15, 18, 19, 33, 40, 44, 48, 63–6, 80, 82, 83, 104, 118
Profit, 5, 66, 80
Programming, 34, 36, 41–4, 49, 50, 53, 55, 56, 60, 89, 92, 113
Projection, 43, 74
Property Rights, 27
Public Sector, 9, 10, 21, 27, 29–31, 69, 70

Qingdao, 24, 124, 130
Quadratic Programming, 53
Quay(s), 16, 17, 70, 83–6, 89, 92
Questionnaire, 86

Rail, 19, 23, 28, 85
Random, 11, 36, 38, 51, 54–6, 60, 61, 75, 89, 106–8, 117, 122
Random Effects, 11, 38, 106, 107, 117, 122
Rankings, 71, 94, 101, 102, 106
Ratio, 4–7, 42, 44, 99
Ray, 6, 50
Reach Stackers, 85
Regional, 9, 26, 88
Regression, 53, 54
Regressors, 57–9, 107
Regulation(s), 21, 25, 26, 80, 143
Regulatory Agencies, 16
Regulatory Function, 28, 29
Reliability, 82
Remuneration, 84
Residual(s), 53–5, 59, 79, 98–100, 107
Road, 19, 28
Robustness, 12, 65, 69, 74, 75, 106
Rotterdam, 24, 87, 124, 130
Routes, 8, 20
Routing, 23

Salaries, 84
Sample, 33, 36, 48, 59, 68, 69–72, 77, 86, 90, 92–5, 97–9, 103–7, 113, 117, 119, 122, 126, 128, 129, 136, 142, 145, 146
Sampling Frame, 68
Scale Efficiency, 5, 94
Scheduling, 19
Sea, 1, 14, 16, 19, 28
Seaport, 68, 69, 124, 137
Service Port, 27
Services, 2, 15, 17–20, 23, 27, 28, 79, 82, 84, 88
Shadow Prices, 44
Shanghai, 24, 124, 129, 130
Shenzhen, 90, 95, 124, 137
Ship Owners, 24, 83
Shippers, 1, 9, 16, 130
Shipping, 1, 8, 9, 16, 17, 19, 20, 25, 64, 69, 80, 88
Shipping Lines, 1, 8, 9, 16, 17, 19, 20, 25, 40, 64, 80, 88, 131, 134, 135
Ships, 8, 14, 16, 17, 20, 25, 64, 83
Simulation, 61
Singapore, 10, 22, 69, 88, 124, 126, 130, 131
Skewness, 90, 91, 94, 99, 101, 107
Slope, 6, 34, 40, 54
Southeast Asia(n), 22, 124
Spanish, 67, 70, 71
Spearman's Rank Order Correlation, 93–5, 101–3, 106, 108–10, 113, 117–19, 122, 130
Specifications, 36, 63, 89, 92, 142
Speed, 8, 83
Stack(ing), 16, 17, 87
Staff, 24, 86
Standard Deviation(s), 90, 91, 94, 101, 104, 119, 136
Standard Error, 90, 91, 94, 101
Standardisation, 1, 18
State-of-the-Art, 8, 80, 81, 112, 145
Statistical Noise, 43, 56, 58, 60, 89, 145
Stevedore(s), 40, 64, 85, 86, 87
Stochastic Frontier Analysis (SFA), 10, 11, 36, 55, 56, 69–74, 93, 98, 102

Storage, 14, 16, 17
Straddle Carrier(s), 85, 89, 92
Strategy, 15, 40, 52, 88, 106
Subset(s), 45, 46, 49
Subsidies, 21, 25
Superstructure, 14, 27, 28
Supply Chain Management, 18, 82, 140
Survey, 73, 86
System(s), 3, 8, 9, 16, 18, 19, 23, 25, 26, 34, 73

TCT Tollerort, 97, 124, 137
Technical Efficiency, 4, 6–8, 12, 48, 53, 59, 94, 141, 145
Technological Innovation, 7, 103, 110, 117, 145
Technolog(y)ies, 4, 6, 9, 17, 18, 25, 36, 48, 49, 52–4, 71, 81, 87, 104, 112, 145
Terminal(s), 2, 9, 16, 17, 19, 24, 29–31, 40, 65, 69–73, 81–92, 94–8, 100, 101, 104, 106, 117, 118, 130, 131, 136, 137, 139, 143, 144
Terminal Operations, 16
Terminal Production, 84, 85, 92
TEUs, 16, 66, 95, 97
Throughput, 10, 40, 63, 65–8, 70, 72, 80, 81, 83, 86–8, 90–2, 95–7, 112, 118, 119
Tianjin, 24, 124, 130
Time-Invariant, 38, 59, 60, 63, 89, 106, 110, 114
Time-Varying, 38, 59–61, 89, 109, 110, 114
Tonnes, 1, 85, 86
Total Factor Productivity, 4, 10
Tracking, 19, 117
Trade, 1, 9, 14, 21, 22, 88
Transit, 14, 20, 23, 51, 65, 66, 82
Translog, 63, 70, 71, 74
Transport(ation), 1, 14–16, 18–20, 23, 26, 64, 66, 80, 82, 140
Transshipment, 16
Truncated Normal, 100
Trust, 69
Turnaround Time, 1, 8, 83

United Kingdom (U.K.), 25, 30, 69, 70, 72, 124
Unbiased, 54, 55, 111
UNCTAD, 1, 14, 15, 64, 84
United States, 18, 23

Validity, 12, 69, 106, 109, 117, 138
Variable Returns to Scale, 42, 43, 50, 73, 74, 94, 103
Variable(s), 12, 44, 45, 51, 56, 60, 66, 68, 73, 77–80, 82–5, 89, 92, 99, 106, 142
Variance, 58, 94, 101
Vector, 50, 53

Vessel, 17, 18, 64
Virtual, 39, 42, 44

Wages, 84
Waste, 14, 26, 48, 73, 88
Window Analysis, 46–8, 103–5, 114, 117–19, 144
Window Width, 46–8, 103, 104, 114
Workers, 70, 86

χ^2, 15, 96, 100, 108, 109, 133

Yard(s), 16–18, 84–6, 89, 92
Yard Gantry Cranes, 85, 86, 89, 92